Neuroscience in Animals

The Neural Basis of Animal Behaviour

Animal Science, Issues and Research Series

Microbiomes in Animal Nutrition, Health, and Disease
Tanmoy Rana, PhD (Editor)
ISBN: 979-8-89530-425-9
eBook ISBN: 979-8-89530-516-4

Snakes: Morphology, Function, and Ecology
David Penning, PhD (Editor)
ISBN: 979-8-88697-855-1
eBook ISBN: 979-8-88697-921-3

Burrow Pattern in Rodents
Chanchal Kumar Manna, PhD
and D. Chattopadhyay, PhD (Editors)
ISBN: 978-1-68507-017-5
eBook ISBN: 978-1-68507-347-3

Nonhuman Primate Models in Preclinical Research. Volume 1: Basics and Regulatory Principles
Huifang Chen, MD, PhD
and Jan A.M. Langermans, PhD
(Editors)
ISBN: 978-1-53619-440-1
eBook ISBN: 978-1-53619-922-2

Nonhuman Primate Models in Preclinical Research. Volume 2: Disease Models
Huifang Chen, MD, PhD
and Jan A.M. Langermans, PhD
(Editors)
ISBN: 978-1-53619-914-7
eBook ISBN: 978-1-53619-931-4

Critical Research Techniques in Animal and Habitat Ecology: International Examples
Kaushalendra Kumar Jha, PhD
and Michael O'Neal Campbell, PhD
(Editors)
ISBN: 978-1-53619-846-1
eBook ISBN: 978-1-53619-859-1

More information about this series can be found at
https://novapublishers.com/product-category/series/animal-science-issues-and-research/

Tantri Dyah Whidi Palupi
Soeharsono
Yeni Dhamayanti
Gracia Angelina Hendarti

Neuroscience in Animals

The Neural Basis of Animal Behaviour

DOI: https://doi.org/10.52305/KKME7835

NOTICE TO THE READER

Library of Congress Cataloging-in-Publication Data

ISBN: 979-8-89530-988-9 (Softcover)
ISBN: 979-8-90134-047-9 (eBook)

Published by Nova Science Publishers, Inc. † New York

Contents

Preface

The study of the neural basis of animal behavior is a profound journey that bridges the complexity of biology with the mysteries of the mind. Neuroscience in Animals: The Neural Basis of Animal Behaviour aims to unravel how the intricate structures and functions of the brain contribute to the behaviors exhibited by animals.

This book begins by tracing the origins of neuroscience, offering a look back at the early foundations of brain science and how it has evolved to inform our current understanding. It delves into the microscopic world of neurons and synapses, exploring their roles in transmitting signals and processing information. From there, we move to the structural organization of the brain, uncovering the complex networks that facilitate thought, perception, and response.

As we explore sensory systems, the text reveals how animals interact with and interpret their environments through the brain's sophisticated sensory networks. The book also covers the concept of neuroplasticity, showing how neural connections adapt and change over time, affecting behavior and learning. The realm of behavioral neuroscience is thoroughly examined, illustrating how brain mechanisms drive actions, social interactions, and decision-making.

The exploration extends to the autonomic nervous system, detailing its pivotal role in maintaining homeostasis and responding to environmental stimuli. Equally important is the neuroendocrine system, which connects neural function to hormonal regulation, affecting everything from stress responses to reproduction. The investigation of cognitive functions highlights how perception, memory, and problem-solving are interconnected with neural circuits, offering insight into animal intelligence and behavior.

Pain perception and nociception are discussed not just from a biological standpoint but with consideration for ethical implications, emphasizing the importance of understanding animal welfare. Comparative neuroscience is woven throughout, demonstrating the diversity and evolutionary significance of neural adaptations across species.

Neuroscience in Animals is an invitation to appreciate the wonders of the animal brain and to deepen our understanding of how different organisms navigate the world. It is a tribute to the scientists and researchers who have dedicated their work to unveiling the mysteries of animal behavior and brain function.

Whether you are a student of the biological sciences, a professional in the field, or simply an enthusiast, this book will expand your understanding of the complex and fascinating ways that the brain shapes the behavior of animals.

Tantri Dyah Whidi Palupi
Faculty of Veterinary Medicine
Universitas Airlangga

Chapter 1

History of Neuroscience – The Science Unveiled: A Journey through Time

The history of neuroscience is a complex and fascinating journey that spans centuries, reflecting humanity's evolving understanding of the nervous system. From the earliest explorations of brain function in ancient civilisations to the sophisticated studies of today, the field has been shaped by groundbreaking discoveries and the contributions of remarkable individuals. Key milestones, such as the identification of neurons, the mapping of brain regions, and the development of modern imaging technologies, have revolutionised our comprehension of how the brain and nervous system operate.

In addition to its scientific advancements, the history of neuroscience also underscores the importance of integrating past insights with contemporary research. By revisiting historical breakthroughs and exploring the perspectives of pioneering figures, modern scientists continue to build on the foundation laid by earlier generations, fostering a deeper appreciation for the intricate workings of the brain and its profound impact on human health and behaviour.

1.1. Early Foundations

1.1.1. Ancient Beginnings

1.1.1.1. Trepanation and Early Surgical Practices

Trepanation, a surgical procedure dating back to the Neolithic period, involved drilling holes into the skull, possibly to treat head injuries or release perceived evil spirits. This practice has been studied extensively, as documented by Lee (2019) in a review of the history of brain research. Evidence of trepanation includes the discovery of several artificially perforated skulls from various locations and periods, particularly the Neolithic era. Interest in these artefacts grew when Ephraim George Squire (1821–1888), an American archaeologist, introduced "Inca skulls" with artificial perforations in New York.

Scholars hypothesise that trepanation was performed to address conditions such as headaches, epilepsy, or mental disorders. The belief that drilling into the skull could release trapped evil spirits may have been inspired by observations of smoke rising, leading to the assumption that spirits could also escape in a similar manner. Remarkably, the practice persisted into the 20th century in some parts of Africa and along the Pacific coast, where it was used to treat epilepsy, headaches, and mental illnesses. Paul Broca, a renowned physician and anthropologist, contributed to the systematic study of trepanation, suggesting that the procedure aimed to provide an outlet for harmful spirits.

Earlier findings, such as the first recorded trepanned skull discovered in 1685 by Bernard de Montfaucon (1655–1741), were initially overlooked, with later theories by researchers like Gaa (1824–1880) and Victor Horsley (1857–1916) gaining more recognition. Beyond its spiritual implications, trepanation has also been interpreted as an early form of advanced surgical intervention, potentially used to treat subdural haemorrhage and other cranial injuries.

1.1.1.2. Edwin Smith Surgical Papyrus

The Edwin Smith Surgical Papyrus stands as one of ancient Egypt's most significant medical texts, offering a comprehensive account of early medical practices and surgical procedures. Dated to approximately 1650-1550 BC, it is regarded as the oldest known surgical treatise, providing valuable insight into the medical knowledge and practices of that era.

Discovered in 1862 near Luxor, Egypt, the papyrus is believed to be a copy of an even older manuscript dating back to around 3000-2500 BC. Edwin Smith, an American Egyptologist, acquired it, and it was subsequently translated and published by James Henry Breasted in 1930.

The document is organised into 48 cases that primarily address traumatic injuries, systematically arranged from head injuries down through the body, mirroring modern anatomical treatises. Each case includes a title, diagnostic procedures, clinical signs, and treatment options, along with conclusions regarding the treatability of the condition. Attributed to Imhotep, this ancient Egyptian text encompasses 48 cases of head injuries and their associated symptoms. It is particularly relevant to neurosurgeons due to its detailed descriptions of head and spine injuries. The papyrus also contains the first recorded instances of various neurosurgical terms. It employs an analytical approach to these cases, classifying them based on prognosis, making it one of the earliest known medical documents related to the field of neuroscience.

1.1.1.3. Hippocratic Contributions

Hippocrates, often referred to as the "father of medicine," emphasised that diseases have natural causes. He was among the first to propose that the brain is the organ responsible for intelligence, reasoning, and conscious life—a perspective that physicians and philosophers in Hellenistic Alexandria further developed. The School of Alexandria, inspired by Hippocratic principles, made significant strides in understanding the anatomy and physiology of the nervous system. These observations were pivotal in deepening our comprehension of the brain's structure and function.

His treatise "On the Sacred Disease" was particularly significant for the understanding of epilepsy and other neurological conditions, including apoplexy, spondylitis, hemiplegia, and paraplegia. Hippocrates revolutionised medical practice by emphasising the importance of thorough clinical evaluation for cranial and spinal disorders. He paired this approach with a humanistic and ethical perspective, placing a strong emphasis on the individuality of each patient. This holistic approach has left a lasting impact on the field of neurosurgery.

Hippocrates' comprehensive studies on head injuries and spinal deformities laid the groundwork for many fundamental neurosurgical principles that remain in use today. His methods and observations inspired subsequent generations to adopt a more systematic and ethical approach to treating neurological disorders.

The Hippocratic Corpus, a collection of approximately 70 medical texts, includes 42 authentic clinical cases and has served as a vital source of medical knowledge for centuries. Hippocrates' emphasis on natural causes for diseases, as opposed to attributing them to divine intervention, marked a significant shift in medical philosophy and practice.

Hippocrates' contributions heralded a golden age in Greek medicine, during which philosophical approaches began to integrate more fully with medical practices. His focus on direct observation and experimentation as the foundation of scientific knowledge remains a cornerstone of modern science. This methodology was further advanced by later scientists such as Herophilus and Erasistratus, who made remarkable progress in the fields of anatomy and physiology.

1.1.2. Classical and Medieval Advances

1.1.2.1. Islamic Golden Age

Following the expansion of Islam, scholars in Al-Andalus (present-day Spain) and other Islamic regions made significant contributions to the field of neuroscience. They translated and expanded upon Greco-Roman medical texts, leading to advancements in neurosurgery, neuropharmacology, and the understanding of neurological diseases.

Abū Bakr Rabi' ibn Ahmad Akhawayni Bukhāri, known as Akhawayni, was a prominent neuropsychiatrist during the early medieval era. His work, "*Hidāyat al-Muta'allimin fi al-Tibb*" (The Students' Handbook of Medicine), is recognised as the first medical textbook in Persian after the advent of Islam. Akhawayni was the first to describe conditions such as sleep paralysis and proposed pragmatic treatments over supernatural explanations. His detailed descriptions of various neuropsychiatric conditions, including meningitis, mania, psychosis, and dementia, align closely with modern medical concepts. Other notable figures include Abulcasis, Averroes, Haly Abbas, Rhazes, and Avicenna (Ibn Sina), among others.

1.1.2.2. Influence on European Medicine

The evolution of European medicine, particularly in the neurosciences, reflects a fascinating interplay of historical and cultural influences. The Islamic Golden Age in Al-Andalus, the rigorous training of Russian neurologists in Western Europe, and the transformative contributions of figures like Andreas Vesalius each represent pivotal chapters in this story.

During the Middle Ages, Al-Andalus became a vibrant centre of medical scholarship, where Islamic scientists meticulously translated, critiqued, and expanded upon Greco-Roman medical texts. Groundbreaking contributions to the field of neuroscience marked this intellectual flourishing. Abulcasis, for instance, introduced surgical methods that remain foundational, while Averroes provided an early description of Parkinson's disease. Avenzoar made significant strides in understanding neurological conditions such as meningitis and explored early neuropharmacology. These advancements underscore the profound impact of Islamic scholars on the broader medical landscape of their time.

In the 19th century, from around the 1850s to 1900, Russian neurologists and psychiatrists shaped their disciplines by engaging deeply with the medical traditions of Western Europe. Many were trained in prestigious institutions in Paris, Berlin, Leipzig, and Vienna, learning from luminaries such as Charcot and Flechsig. This cross-cultural exchange facilitated the early establishment of neurology and psychiatry as distinct clinical fields in Russia, laying the groundwork for advances that paralleled or even preceded developments in other parts of Europe.

Equally transformative was the work of Andreas Vesalius, whose empirical approach to anatomy redefined the study of the human body. By meticulously dissecting cadavers and challenging the dogmatic adherence to ancient texts, Vesalius laid the groundwork for modern neuroanatomy. His seminal work, *De humani corporis fabrica*, not only corrected prevailing misconceptions but also inspired a more scientific approach to medicine that reverberated through subsequent centuries.

Vienna, too, holds a special place in the history of neurological science. As the home of Sigmund Freud, the city became synonymous with the early development of psychoanalysis and the integration of psychiatry with neurology. Even today, Vienna remains a beacon for neurological research and education, hosting international conferences and fostering collaborative advancements in both clinical and experimental fields.

1.1.2.3. Renaissance Breakthrough

The field of neuroscience is undergoing a remarkable resurgence, blending historical methodologies with the breakthroughs of modern technology. This renewed momentum is driven by the incorporation of dynamical systems theory, next-generation molecular tools, and advanced imaging techniques, all of which are redefining our understanding of neural processes and brain anatomy. Interestingly, the application of dynamical systems theory in neuroscience, whilst enjoying a revival, is rooted in foundational works such as the Hodgkin-Huxley and FitzHugh-Nagumo models. These early frameworks offered critical insights into neural dynamics, particularly in motor control, and have influenced contemporary perspectives that view the brain as a dynamic system rather than merely a representational one.

Historically, significant strides in neuroscience can be traced back to the Renaissance, where pioneers like Leonardo da Vinci and Andreas Vesalius laid the groundwork for modern anatomical science. Leonardo da Vinci's anatomical studies marked a pivotal shift from speculative theories to empirical observation. Initially influenced by medieval ideas, his later sketches, based on dissections, provided more accurate representations of the brain, challenging the long-held belief that mental functions were confined to the ventricles.

Building on this shift, Andreas Vesalius advanced the field further in the 16th century by prioritising human dissections over the animal-based studies championed by Galen for centuries. Vesalius's meticulous approach culminated in the publication of *De Humani Corporis Fabrica* in 1543, a landmark text renowned for its scientific precision and artistic excellence. Although Vesalius did not fully unravel the functional intricacies of the brain, his detailed anatomical illustrations established a foundation for future exploration, emphasising the importance of direct observation in studying the nervous system. This Renaissance legacy continues to inspire the modern neuroscientific endeavour, bridging centuries of discovery with the promise of new horizons.

1.1.3. The Neuron Doctrine

The neuron doctrine, formulated in the late 19th century, represents a pivotal milestone in neuroscience, proposing that the neuron is the fundamental structural and functional unit of the nervous system. This principle has profoundly shaped our understanding of the organisation, functionality, and development of the nervous system.

The emergence of this doctrine is credited to the groundbreaking work of scientists such as Santiago Ramón y Cajal and Camillo Golgi, whose conflicting views ultimately fueled progress in the field. While Golgi's reticular theory argued for a continuous network of nerve fibres, Cajal's meticulous observations demonstrated that neurons are discrete entities. Their combined efforts earned them the Nobel Prize in Physiology or Medicine in 1906. Notably, the term "neuron" was introduced by Vilhelm von Waldeyer in 1891, inspired by the Greek word for "sinew," reflecting the structural and functional continuity that neurons provide within the nervous system.

Central to the neuron doctrine are three fundamental principles: neurons serve as the genetic, anatomical, functional, and trophic units of the nervous system; each neuron originates from a single neuroblast and maintains its processes through its cell body; and neurons communicate via specialised connections called synapses. These concepts provided the foundation for our understanding of neural architecture and signalling.

However, as neuroscience has advanced, the classical neuron doctrine has faced re-evaluation and refinement. Emerging evidence reveals that the traditional view of neurons as isolated functional units is oversimplified. Recent findings highlight the role of intercellular communication mechanisms, such as gap junctions and dendritic action potentials, in neural processing. Neuromodulation and extrasynaptic neurotransmitter release have also been shown to influence brain activity significantly. Furthermore, dendrodendritic synapses, which are particularly prevalent in mammalian brains, suggest that functional units may extend beyond individual neurons, indicating a more interconnected and dynamic system than previously envisioned.

These developments underscore the evolving nature of neuroscience, prompting researchers to refine and expand foundational theories as our understanding of the field continues to deepen. For a junior lecturer with a passion for the intricate history and ongoing discoveries in the field, the neuron doctrine serves as a testament to the balance between foundational principles and the necessity of scientific curiosity and adaptability.

1.2. The Emergence of Modern Neuroscience

The emergence of neuroscience as a distinct field during the late 1950s and early 1960s marked a transformative period in the biological sciences. This new discipline unified diverse areas of study—neurophysiology, neuroanatomy, neurochemistry, and behavioural research—into a cohesive framework, driven by the efforts of influential pioneers such as David McKenzie Rioch, Francis O. Schmitt, and Stephen W. Kuffler.

1.2.1. Foundational Figures

1.2.1.1. David McKenzie Rioch, Francis O. Schmitt, and Stephen W. Kuffler

David McKenzie Rioch played a crucial role in integrating various subfields of neuroscience, fostering collaboration among researchers from different disciplines. His efforts helped to establish neuroscience as a cohesive field of study. Schmitt was instrumental in applying molecular and cellular biology techniques to the study of the nervous system. His work significantly advanced our understanding of neuronal signalling and neural development, which are fundamental to modern neuroscience. Kuffler is particularly noted for his pioneering research in neurophysiology and neuroanatomy. His contributions were pivotal in elucidating the function of major sensory and motor systems in the brain, thereby shaping the direction of contemporary neuroscience research.

1.2.1.2. Santiago Ramón y Cajal and Camillo Golgi

The contributions of Santiago Ramón y Cajal and Camillo Golgi are foundational to modern neuroscience. Their pioneering work on the structure of the nervous system earned them the Nobel Prize in Physiology or Medicine in 1906, marking the beginning of a new era in the field of neuroscience.

1.2.1.2.1. Golgi's Black Reaction

Camillo Golgi developed the Black Reaction (*La reazione nera*), a staining technique that allowed for the detailed visualisation of neural structures, including axons and dendrites. This innovation revealed the intricate architecture of the nervous system and provided an essential tool for neuroanatomical research, paving the way for subsequent discoveries.

1.2.1.2.2. Cajal's Neuron Doctrine

Building on Golgi's discoveries, Santiago Ramón y Cajal proposed the Neuron Doctrine, establishing the neuron as the fundamental structural and functional unit of the nervous system. He introduced the concept of dynamic polarisation, which describes the directional flow of information within neurons, and meticulously mapped neural circuits. His insights into brain plasticity—the brain's ability to form new neural connections—remain integral to modern research on learning, memory, and recovery from neurological damage.

Cajal's legacy extended through his students, such as Pío del Río-Hortega, who identified oligodendrocytes and microglia, and Fernando de Castro, who described the innervation of blood vessels and identified chemoreceptors in the carotid body. Together, their contributions continue to influence the ever-evolving field of neuroscience.

The collective efforts of these figures laid the intellectual groundwork for neuroscience, demonstrating how a combination of innovation, collaboration, and detailed observation can unravel the complexities of the nervous system. This historical foundation continues to inspire and inform the pursuit of discoveries in the field today.

1.2.2. Technological and Methodological Advances

The rapid evolution of neuroscience has been significantly propelled by technological and methodological breakthroughs, which have transformed our understanding of the brain's complexity. These innovations have enabled researchers to explore neural structures and functions with unprecedented precision, offering new insights into both fundamental and clinical aspects of neuroscience.

1.2.2.1. Key Technologies Advances

The emergence of cutting-edge technologies has been instrumental in advancing neuroscience research. Recent progress in neurotechnologies, such as electrical, optical, and microfluidic neural interfaces, has enhanced the ability to probe and manipulate neural circuits, ranging from individual neurons to entire brain networks. These tools, increasingly adopted in studies using animal models, are shaping the future of neuroscience. Neuroimaging technologies, psychopharmaceutical advancements, and brain stimulation techniques have not only revolutionised the capacity to explore brain function but have also raised profound ethical and social questions, challenging traditional perspectives on selfhood and the brain-body relationship. The growing emphasis on large-scale collaborations and open science initiatives further democratizes access to neural data, fostering inclusivity in scientific exploration.

Among these innovations, optogenetics stands out as a transformative tool, enabling researchers to control neuronal activity with light, thus offering precise insights into brain function and behaviour. Similarly, advanced imaging methods, such as high-resolution EEG and calcium imaging, provide dynamic, real-time visualisations of neural activity, complementing traditional static imaging approaches. Technologies like multielectrode arrays and optogenetics also support multineuronal recording, yielding high-resolution data on neural dynamics and connectivity across networks.

1.2.2.2. Methodological Advances

Alongside technological advancements, methodological progress has redefined how researchers analyse and interpret neural data. The integration of large-scale datasets with machine learning algorithms has revolutionised data analysis, making it possible to decode brain-behaviour relationships with remarkable accuracy. Genomic and behavioural approaches are further enriching neuroscience, enabling a deeper understanding of the genetic underpinnings of behaviour and the pathophysiology of neurological disorders. Meanwhile, multi-task learning (MTL), a machine learning approach that integrates diverse data modalities, is opening new avenues for translational research by unravelling the complex mechanisms underlying neurodegenerative and psychiatric diseases.

Together, these technological and methodological advancements not only expand our capacity to explore the brain but also redefine the questions we ask about its functions and dysfunctions. They represent a dynamic and ever-evolving toolkit, empowering researchers to bridge the gap between fundamental neuroscience and real-world applications. For a junior academic immersed in the field, these developments highlight the immense potential of interdisciplinary approaches and the exciting possibilities that lie ahead in unravelling the mysteries of the human brain.

1.2.3. The Integration of Disciplines

The evolution of modern neuroscience owes much to the seamless integration of previously distinct disciplines into a unified field of study. This process has profoundly enhanced our understanding of the brain and its functions. This interdisciplinary transformation unfolded in several phases, each contributing uniquely to the development of neuroscience as it is known today.

During the initial phase in the 1950s and 1960s, traditional fields such as neuroanatomy, neurochemistry, neuropharmacology, and neurophysiology began to converge, laying the groundwork for modern neuroscience. This foundational integration allowed researchers to examine the brain from multiple perspectives, facilitating a deeper understanding of its structural and functional complexities. The subsequent incorporation of molecular biology and genetics marked a significant leap forward in the field. By focusing on the genetic underpinnings of neurological disorders, scientists were able to investigate these conditions at a molecular level, even in the absence of prior knowledge about their biochemical mechanisms. This approach provided valuable insights into how genetic variations contribute to the development and progression of neurological and psychiatric diseases.

The mid-1980s marked another transformative phase, characterised by the convergence of cognitive psychology and neuroscience, which gave rise to cognitive neuroscience. This new subfield aimed to unravel the neural mechanisms underlying cognitive functions, such as memory, attention, and decision-making, thereby bridging the gap between brain structure and behaviour. Cognitive neuroscience has not only expanded our understanding of mental processes but also laid the foundation for practical applications in fields such as education, artificial intelligence, and clinical therapy.

Despite these advancements, challenges persist, particularly in uncovering the complex causes of neurological and psychiatric disorders. While significant strides have been made in mapping the genetic basis of these conditions, much remains to be understood about the intricate interplay between genes, neural circuits, and environmental factors. Emerging approaches, such as integrating network science with neuroscience, offer promising opportunities to decode the brain's intricate networks and develop innovative therapeutic strategies tailored to specific neural circuits. This ongoing journey highlights how interdisciplinary collaboration and innovation continue to drive the field forward, promising a deeper understanding of the brain and its vast potential.

1.3. Major Revolutions in Neuroscience

The history of neuroscience is marked by transformative milestones that have continuously reshaped our understanding of the nervous system, dating back to as early as the 6th century BCE. These revolutions have not only challenged fundamental beliefs about how the brain works but have also redefined the methods and strategies used to study it. Unlike the paradigm shifts described by Thomas Kuhn, where competing theories or anomalies drive scientific revolutions, recent breakthroughs in neuroscience have been fuelled by the invention of novel experimental tools. For instance, the development of genetically engineered animal models has revolutionised the study of cognitive processes. At the same time, cutting-edge techniques such as optogenetics and designer receptors exclusively activated by designer drugs (DREADDs) have enabled the precise manipulation of neural circuits, opening new frontiers in brain research.

In addition to experimental advances, the integration of genetics and genomics has catalysed a shift towards discovery-based research, complementing traditional hypothesis-driven methods in molecular, cellular, and systems neuroscience. The availability of large-scale genetic and phenotypic data sets, combined with sophisticated tools for data integration and analysis, has provided more profound insights into neuronal diversity and the molecular underpinnings of neurological diseases. These advancements not only enhance our understanding of brain function but also pave the way for potential therapeutic interventions.

Simultaneously, theoretical neuroscience has undergone rapid growth over the past two decades, attracting researchers from diverse disciplines, including physics, mathematics, computer science, and engineering. This interdisciplinary approach has injected fresh perspectives and innovative methodologies into the field, significantly influencing the trajectory of modern neuroscience. Together, these revolutions illustrate the dynamic and evolving nature of neuroscience, a field that continually pushes the boundaries of our knowledge about the brain and its intricate workings.

1.4. History of Animal Neuroscience

The history of animal neuroscience spans over 25 centuries, reflecting the evolution of ideas, methodologies, and technologies that have deepened our understanding of the nervous system. These narrative highlights pivotal moments and ongoing transformations, emphasising how historical advances have shaped modern neuroscience.

1.4.1. Foundational Developments in Early Animal Neuroscience

Animal neuroscience began with philosophical speculation and rudimentary anatomical studies, gradually evolving into systematic scientific inquiry. The earliest records date back to approximately 500 BCE, during which ancient philosophers, such as Aristotle, proposed theories about cognition, the soul, and the body. This period was characterised by reductionist studies involving invasive *in vivo* and *ex vivo* examinations of human and animal brains. These efforts, though limited by the tools of the time, laid the groundwork for understanding the brain as a distinct organ linked to thought and behaviour.

During the 16th to 18th centuries, comparative anatomy gained prominence, as researchers such as Giovanni Battista Morgagni utilised animal evidence to study brain diseases. This period saw the integration of natural philosophy with medical theories, exemplified by the study of birdsong to explore the connections between language and cognition. Philosophical debates about the mind-body relationship and the nature of the soul significantly enriched the intellectual landscape, although empirical investigation was still in its infancy.

1.4.2. Second Era: The Systematic Experimentation and Technological Advances (17th Century)

The 17th century marked a turning point with the introduction of invasive methods for studying living tissues. Early modern philosophers and anatomists began challenging Cartesian dualism, advocating for a more integrated view of the mind and body. This period was characterised by systematic experimentation that blended philosophical ideas with medical practices, laying the foundation for neuroscience as a scientific discipline.

The advent of microscopes in the 19th century heralded a revolution in neuroanatomy. Scholars such as Robert Remak, Gustav Valentin, and Jan Evangelista Purkyně employed histological methods to observe nerve tissue in unprecedented detail. Although these researchers did not yet identify neurons as discrete entities, their work significantly advanced the understanding of nervous system structures, influencing future breakthroughs.

1.4.3. Third Era: Cellular Neurophysiology and Model Organisms (1838–Present)

The discovery of neurons as distinct cellular units by Camillo Golgi and Santiago Ramón y Cajal in the late 19th century marked the beginning of the era of cellular neurophysiology. This era, spanning from 1838 to the present, has focused on the cellular mechanisms of neural function, facilitated by advancements in staining techniques and electrophysiology.

The 20th century saw the emergence of model organisms, particularly rodents, as dominant subjects in neuroscience research. This shift was driven by the practicality of maintaining these species, their physiological similarities to humans, and the standardisation of experimental protocols. However, this reliance on a narrow range of species has sparked debates about the need for diversity in animal models. Historical milestones, such as the study of squid giant axons and the identification of critical motor control principles in invertebrates, underscore the value of exploring non-traditional models to uncover conserved mechanisms and novel insights.

1.4.4. Fourth Era: The Age of Non-Invasive, Holistic Neuroscience

The current era, characterised by non-invasive, multimodal approaches, represents a paradigm shift in animal neuroscience. Advances in neuroimaging, optical imaging, and electrophysiology now enable researchers to study awake, behaving animals in real-time, providing a holistic understanding of brain function.

1.4.4.1. Technological Innovations in Neuroscience

Modern methods such as two-photon imaging, optogenetics, and functional MRI (fMRI) have revolutionised how researchers investigate the brain. These techniques enable the simultaneous recording of neural activity and behaviour, revealing dynamic neural circuits underlying cognition and emotion. For instance, optical imaging in freely moving animals has facilitated the discovery of place, grid, and head-direction cells, which are essential for spatial navigation.

1.4.4.2. Multimodal and Long-Term Recordings

The integration of multimodal imaging and long-term neuronal recordings has opened new avenues for studying neural plasticity, memory, and decision-making. Advances in electrode design and surgical techniques have minimised tissue damage, enabling chronic studies that provide insights into brain-behaviour relationships over time.

1.4.5. Challenges and Calls for Diversity in Neuroscience

While significant progress has been made, challenges persist in the field of animal neuroscience. Ethical considerations, such as minimising animal suffering, and methodological issues, such as standardising imaging protocols across species, remain pressing concerns.

1.4.5.1. Expanding Beyond Rodents

The historical reliance on rodents has been instrumental but also limiting. Researchers increasingly advocate for using a broader range of species, including non-model organisms such as cephalopods, zebrafish, and large-brained farm animals. These species offer unique opportunities to explore evolutionary adaptations, behavioural nuances, and conserved neural mechanisms that are difficult to study in traditional models.

1.4.5.2. Behavioural and Ecological Contexts

Incorporating species-specific behaviours and ecological contexts into experimental designs is essential for improving the translational validity of animal models. Ethoexperimental approaches, which align experimental conditions with naturalistic behaviours, provide a more accurate understanding of how neural circuits operate in real-world scenarios.

1.4.6. Ethical Considerations and Future Directions

Ethical concerns have long been a cornerstone of animal neuroscience. The shift towards non-invasive techniques and the exploration of alternative models reflects a commitment to reducing animal suffering while advancing scientific inquiry. Innovations such as miniature microscopes and portable imaging systems hold promise for studying neural processes in ecologically relevant settings.

Looking ahead, the integration of artificial intelligence and machine learning into data analysis will enhance the interpretation of complex neural datasets. Standardisation of methodologies and the development of cross-species databases will further facilitate comparative studies, bridging gaps between neuroscience, behaviour, and evolution.

1.5. Integration with Clinical Disciplines

The integration of neuroscience with clinical disciplines has evolved significantly over time, driven by shared advancements and a growing recognition of their interdependence. This merging aims to enrich the understanding and treatment of mental health and neurological disorders by synthesising insights from neurobiology with clinical expertise. The relationship between neuroscience and clinical practices dates back to early collaborations that laid the foundation for modern interdisciplinary approaches. Historical documents, research findings, and past clinical experiences have all played a pivotal role in shaping current neuroscience theories and practices. For instance, the incorporation of neuroscience into clinical social work has shed light on the neurophysiological underpinnings of personality disorders, defence mechanisms, and attachment styles, underscoring the profound connections between cognitive processes and brain functions. Denmark's history in neuroscience exemplifies how close cooperation between basic and clinical neuroscience, especially in neurology and psychiatry, has facilitated groundbreaking translational research and the development of new therapeutic strategies.

The integration of neuroscience into education has also seen notable advancements. Clinical Neuroscience courses have emerged as a bridge between fundamental research and clinical application, providing a solid foundation in neuroanatomy, neurochemistry, and neural development that can be applied to the understanding and treatment of mental health disorders. Modern medical education has further evolved to incorporate a more integrated approach that combines basic sciences, clinical sciences, and health system studies, promoting a holistic understanding of neuroscience in healthcare. This educational model has been shown to improve both student performance and the practical application of knowledge when addressing neurological cases.

Collaboration across disciplines is another cornerstone of effective integration. Neuroscience research often involves collaborations with psychology, education, and clinical practice, facilitating the translation of basic research findings into practical clinical interventions. The proposed use of a whole-animal approach to neuroscience research exemplifies how integrating basic and clinical research can deepen our understanding of complex neuropsychiatric phenomena, such as fear-conditioned responses. This comprehensive, interdisciplinary approach not only enhances the breadth of scientific inquiry but also promotes the development of more effective, evidence-based clinical treatments.

Conclusion

The historical journey of neuroscience is a testament to humanity's enduring curiosity and determination to unravel the mysteries of the brain and nervous system. From trepanation in prehistoric times and the surgical knowledge of ancient civilisations to the foundational doctrines of the Renaissance and the empirical rigour of modern neuroscience, the field has evolved through a continuous interplay of observation, experimentation, and innovation. The transition from philosophical speculation to evidence-based scientific practice reflects a profound transformation in how we understand cognition, behaviour, and neurological disorders.

Crucially, the history of neuroscience is a rich narrative of interdisciplinary collaboration. Contributions from medicine, psychology, anatomy, pharmacology, and even philosophy have converged to form an integrative discipline capable of addressing both fundamental and clinical questions. The development of animal neuroscience, the refinement of ethical research practices, and the incorporation of emerging technologies underscore the dynamic and adaptive nature of the field.

In the present era, characterised by technological sophistication and cross-disciplinary integration, neuroscience continues to bridge the gap between theory and practice. It empowers researchers and clinicians alike to translate complex neural insights into tangible health interventions. As we look forward, the lessons of history serve not only as a foundation but as a guiding framework, reminding us that progress in neuroscience, as in all sciences, is built upon the shoulders of those who dared to question, explore, and innovate.

Chapter 2

From Sparks to Signals – Inside Neurons and Synapses

Neurones and synapses constitute the foundational elements of the nervous system, underpinning the transmission and integration of information via complex electrochemical signalling. Neurones function as the principal operational units, while synapses, as specialised junctions, facilitate intercellular communication. This interplay forms the basis of higher-order cognitive functions, including learning and memory, and serves as the substrate for various neurodevelopmental and neuropsychiatric disorders.

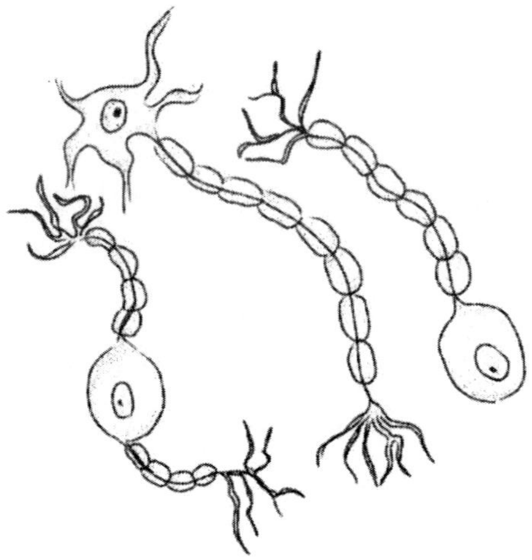

Figure 1. Schematic Figure of three types of neurons (Private documentation, 2025).

2.1. Neurone

Neurones display an extraordinary degree of structural and functional heterogeneity, a feature fundamental to the brain's complexity. This diversity is reflected in their morphology, molecular profile, and functional roles across distinct brain regions.

Compartmentalisation defines neurone structure into soma, dendrites, axons, synaptic terminals, and dendritic spines. These subcellular domains are supported by the cytoskeleton, composed of microtubules, actin filaments, and intermediate filaments. The plastic nature of the cytoskeleton, modulated by actin-binding proteins and tropomyosin isoforms, enables neuritic growth, branching, and synaptic restructuring, processes critical for development and plasticity. Structurally, neurones are categorised as unipolar, bipolar, pseudounipolar, or multipolar. Morphological distinctions, such as those observed between the superior temporal gyrus and the anterior cingulate cortex, align with specific functional specialisations and are frequently disrupted in conditions like schizophrenia.

Functionally, the neocortex epitomises neuronal adaptability, orchestrating complex tasks such as reasoning and perception. Transcriptional regulation during development directs this specialisation, while scRNA-seq techniques have unveiled molecular underpinnings of neuronal identity. The connectome's hierarchical structure integrates neurones into modular frameworks, enabling both localised processing and global integration. Dysfunctions in this architecture, as seen in epilepsy, underscore the significance of structural-functional coherence.

Neurones are broadly classified into projection neurones and interneurones based on connectivity, molecular signatures, and axonal trajectories. Transcriptomic analyses have further refined this taxonomy. For instance, motor neurones, crucial for muscular control, exhibit selective vulnerabilities in disorders such as ALS. Pyramidal and axo-axonic cells in the cortex exemplify how morphology governs synaptic input-output functions. Moreover, differences in epigenomic landscapes, including chromatin accessibility and methylation profiles, contribute to neuronal functional diversity.

2.1.1. Developmental Processes

Neuronal development commences with neural plate induction during embryogenesis, orchestrated by signals like BMPs and Wnts. Neural precursors derived from the ectoderm form the neural tube, which undergoes axial regionalisation.

Neurogenesis entails the proliferation and differentiation of neural stem cells, followed by the migration and positioning of neurones. Axonal pathfinding, governed by cues such as netrins and semaphorins, guides connectivity. Synaptogenesis ensues, shaped by both genetically encoded and activity-dependent mechanisms. Spontaneous early activity promotes maturation and network formation, transitioning from localised activity to coherent circuit-level dynamics.

2.1.2. Functional Roles and Specifications

Neuronal function is dictated by compartment-specific properties such as dendritic ion channel distributions, which govern integration and responsiveness. Inhibitory interneurones, such as those expressing parvalbumin or somatostatin, maintain excitatory-inhibitory balance essential for network stability and learning.

Mechanisms like local mRNA translation and microRNA-mediated regulation facilitate plasticity and adaptive connectivity. Optogenetics and scRNA-seq have empowered precise circuit dissection, enhancing our understanding of behaviourally relevant neuronal activity.

2.1.3. Challenges and Future Directions

Despite advances, fully resolving neuronal heterogeneity and circuit logic remains challenging. Integrative approaches (combining anatomical, genetic, and transcriptomic modalities) are imperative. Computational modelling (e.g., via NEURON) and genetic access to activated circuits offer unprecedented opportunities for dissecting brain function and developing targeted therapies.

This section highlights the dynamic complexity of neurones, from their molecular identity to circuit integration, laying the groundwork for understanding brain function and dysfunction.

2.2. Synapse

Synapses are the specialised contact points through which neurones communicate with one another or with effector cells. They are indispensable for processing information and mediating behaviour, and are central to numerous neuropathologies.

2.2.1. Synapse Structure and Function

Synapses consist of three primary components: the presynaptic terminal, the synaptic cleft, and the postsynaptic specialisation. The presynaptic terminal is responsible for releasing neurotransmitters into the synaptic cleft, where they bind to receptors on the postsynaptic membrane, enabling signal transmission. Synapses can be broadly categorised into chemical and electrical types, each serving distinct but complementary roles in neural communication.

2.2.1.1. Chemical Synapses

Chemical synapses involve the release of neurotransmitters from the presynaptic neurone, which then bind to receptors on the postsynaptic neurone. This type of synapse is highly specialised for rapid and specific signal transmission and is crucial for processes such as learning and memory. Synaptic plasticity, the ability of synapses to strengthen or weaken over time, is a key feature of chemical synapses and underlies many forms of learning and memory.

Chemical synapses use neurotransmitters to activate or inhibit the postsynaptic neurone. This process involves the release of neurotransmitters from synaptic vesicles into the synaptic cleft, where they bind to specific receptors on the postsynaptic membrane, leading to either excitatory or inhibitory postsynaptic potentials.

Chemical synapses exhibit significant plasticity, which is essential for learning and memory. This plasticity is well studied and involves mechanisms such as long-term potentiation (LTP) and long-term depression (LTD).

The concept of chemical synaptic transmission was suggested as early as 1877 and was further developed by researchers such as Sherrington, who coined the term “synapse” in 1897.

2.2.1.2. Electrical Synapses

Electrical synapses, formed by gap junctions, allow direct ionic current flow between neurones. These synapses are known for their role in synchronising neuronal activity and can influence various computational functions, such as spike-dependent inhibition and signal discrimination. Electrical synapses are modifiable and can dynamically regulate neural circuits, adding complexity to neuronal integration and action potential generation. Electrical synapses transmit information through passive transmission of voltage via gap junctions, which are intercellular channels that connect adjacent neurones. This allows for almost instantaneous signal transmission.

Electrical synapses are critical in synchronising neuronal activity. They can enhance the degree of synchronisation in neural networks, especially in the presence of heterogeneous excitability among neurones.

Although less understood than chemical synaptic plasticity, electrical synapses also exhibit forms of plasticity, such as long-term depression (eLTD) and long-term potentiation (eLTP), which are crucial for modulating network activity.

Chemical and electrical synapses often interact within the same neural circuits to regulate communication and network function. For instance, while chemical synapses can suppress network heterogeneity, electrical synapses counterbalance this suppression by enhancing synchronisation. Furthermore, mixed synapses, which combine chemical and electrical components, provide bidirectional communication, enhancing synaptic efficacy in various neural circuits, particularly in the mammalian spinal cord.

2.2.2. Synapse Formation and Remodelling

Synapse formation and remodelling are fundamental processes that underlie learning, memory, and overall neuronal function. These processes include synaptogenesis, which refers to the formation of new synapses, and synaptic plasticity, the modification of synaptic strength and structure in response to various stimuli.

2.2.2.1. Synaptogenesis

Synaptogenesis involves the interaction of cell adhesion molecules (CAMs) and extracellular matrix (ECM) components, which promote synaptic differentiation and maturation. Proteins such as thrombospondins and neuronal pentraxins, along with guidance molecules like morphogens, regulate the number, location, and strength of synapses. Zinc ions also play a critical role by influencing the recruitment of ProSAP/Shank proteins to postsynaptic densities, which are vital for synaptic stability and density.

2.2.2.2. Synaptic Plasticity

Plasticity allows synapses to adapt dynamically to neuronal activity. Mechanisms of plasticity include activity-dependent remodelling, where repetitive stimulation induces stable structural changes, and the influence of neurotrophins and hormones, which modulate receptor expression and activation. Microglia also contribute to synaptic plasticity by engulfing extracellular matrix components, facilitating synaptic remodelling essential for memory consolidation.

2.2.3. Synaptic Dysregulation and Disorders

Synaptic dysregulation is increasingly recognised as a key mechanism underlying neuropsychiatric and neurodevelopmental disorders. Genetic mutations in synaptic proteins, such as disrupted-in-schizophrenia 1 (DISC1) and fragile X mental retardation protein (FMRP), are linked to conditions like autism spectrum disorder (ASD), intellectual disability (ID), and schizophrenia. These mutations often affect excitatory synapse function, leading to abnormalities in dendritic spine density and synaptic transmission.

Environmental factors, such as maternal immune activation (MIA), further contribute to synaptic dysfunction. MIA has been shown to reduce dendritic spine density and alter synaptic connectivity, resulting in behavioural impairments in animal models. The excitatory/inhibitory (E/I) balance hypothesis is central to understanding these disorders, with alterations in receptor densities playing a significant role in their pathogenesis.

Therapeutic strategies targeting synaptic dysregulation include reversing synaptic abnormalities through interventions such as anti-inflammatory treatments and synaptic remodelling therapies. Early intervention during development holds promise, but emerging evidence suggests that adult neuroplasticity could also provide therapeutic windows, offering hope for addressing complex neuropsychiatric conditions.

This section highlights the critical role of synapses in neuronal communication, the adaptability of neural circuits through plasticity, and the devastating effects of synaptic dysregulation on brain function. Understanding these processes is essential for advancing treatments for neurological and psychiatric disorders.

Conclusion

This chapter has explored the foundational architecture and dynamic functionality of neurones and synapses (core components of the nervous system) responsible for the electrochemical communication that underpins all cognitive and behavioural processes. Neurones, with their remarkable diversity in form, function, and molecular identity, constitute the building blocks of neural circuits. Their compartmentalised structures and plastic cytoskeletal elements allow them to perform specialised roles across various brain regions. The developmental pathways of neurones, from embryogenesis to circuit integration, are finely tuned by molecular cues and activity-dependent processes, reflecting the delicate orchestration required for brain maturation and plasticity.

Synapses, as the communication bridges between neurones, are equally complex. The dichotomy between chemical and electrical synapses illustrates the diversity of mechanisms by which information is transmitted, modulated, and integrated. Chemical synapses allow for specific and flexible signalling through neurotransmitter release and receptor binding, while electrical synapses ensure synchronisation and rapid conduction via gap junctions. These systems do not function in isolation; rather, their integration—along with structural and molecular plasticity—forms the substrate for learning, memory, and adaptive behaviour.

Furthermore, this chapter has illuminated the intricate processes of synapse formation and remodelling, governed by molecules such as CAMs, morphogens, and matrix proteins. These processes are not static but evolve in response to activity and experience, enabling synaptic networks to adapt across the lifespan. However, when these processes are dysregulated, whether by genetic mutations or environmental insults, they contribute to the emergence of neurodevelopmental and neuropsychiatric disorders. Understanding the balance between excitation and inhibition, and the plasticity that underlies it, is therefore not only critical for basic neuroscience but also for the development of effective interventions.

In summary, neurones and synapses are not merely structural units, but dynamic entities that collectively shape brain function. Their study continues to provide essential insights into the neural basis of behaviour and cognition, laying the groundwork for addressing complex brain disorders through molecular, cellular, and systems-level understanding.

Chapter 3

Pathways of Thought – A Neuroanatomical Exploration

Neuroanatomy is a critical field within medical education and neuroscience, focusing on the structure and organisation of the nervous system. It is essential for understanding brain function, development, and disease. Recent advancements have introduced ontology-based models to represent neuroanatomical knowledge. These models enable symbolic lookup, logical inference, and mathematical modelling of neural systems, facilitating applications such as surgical planning and personalised care for neurological diseases. The ontology-based approach encodes both structural and functional aspects of neuroanatomy, allowing for computational evaluation and automated reasoning applications.

3.1. Neuroanatomical Terminology

Neuroanatomical terminology is a critical aspect of neuroscience, providing a standardised language for describing the structures and functions of the nervous system. Recent developments and historical influences have shaped the current terminology, impacting both research and clinical practice.

3.1.1. Recent Developments in Neuroanatomical Terminology

3.1.1.1. Terminologia Neuroanatomica

The Terminologia Neuroanatomica (TNA) represents the latest revision of neuroanatomical nomenclature, updating and refining the systems previously established in the Terminologia Anatomica (1998) and the Terminologia Histologica (2008). The TNA integrates a hierarchical and embryologically based classification system for brain structures, emphasising the prosomeric model for the forebrain (prosencephalon). This framework seeks to standardise communication in neuroanatomical research by offering a coherent and contemporary reference system.

The TNA, developed under the Federative International Programme for Anatomical Terminology (FIPAT) of the International Federation of Associations of Anatomists (IFAA), organises terms in six columns: Latin official term, Latin synonyms, English official term (UK spelling), English official term (US spelling), English synonyms, notes on changes, related terms, and eponyms.

UK and US English terms are considered equivalent, with hyphenated forms generally preferred in UK English and non-hyphenated forms in US English. Latin terms are also updated to accommodate synonyms such as dorsalis and ventralis for posterior and anterior.

3.1.1.2. Historical Roots of Anatomical Nomenclature

The practice of naming anatomical structures dates back to Alkmaion of Croton (500–450 BCE), who is credited with distinguishing arteries (ἀρτηρία) from veins (φλέψ) and identifying the optic nerve. Subsequent contributions by Aristotle (384–322 BCE) and the Hippocratic Corpus significantly influenced early anatomical understanding, though these texts often contained inaccuracies due to limited anatomical exploration.

Aristotle is recognised as the first to systematically dissect animals, documenting structures such as muscles (μῦς), bones (περί αρθρων), and nerves (νεῦρον). He introduced terms like ὄργανον (organ), signifying body parts with specific functions, and conceptualised the body's composition as "fabrics" (θηρά). Many terms introduced by Aristotle, such as χολή (bile), μυελός (marrow), and αἷμα (blood), remain foundational in modern anatomical terminology.

The work of Greek physicians in Alexandria, such as Herophilus of Chalcedon (335–280 BCE) and Erasistratus of Keos (310–250 BCE), marked the first documented human dissections. Despite their controversial methods, terms like νεῦρον (nerve) and ἐγκέφαλος (brain) from their writings persist in contemporary terminology.

3.1.1.3. Greek to Latin Transition

Although Greek dominated early anatomical terminology, Aulus Cornelius Celsus (c. 25 BCE – 50 CE) was instrumental in Latinising anatomical terms in his work *De medicina libri octo*. He introduced Latin translations such as *vertebra* and *jejunum*, and adapted Greek terms like *trachea* into Latinised forms. This marked the beginning of a more structured Latin-based anatomical lexicon. Examples of anatomical terms that derive from the original Greek can be seen in Table 1.

Table 1. Examples of anatomical terms (Celsus, 1465)

Latin Term	Greek Original
Arteria	’Αερτερια (Aerteria) -airpipe
Auris	’Ους, ’Οτος (Ous, Otos)
Caecum	τυφλον ’Εντερον (typhlon Enteron)
Caput	Κεφαλη (Kephale)
Carotidas	Καροτιδας (Karotidas)
Cartilago	Χονδρος (Chondros)
Cerebrum	’Ενκεφαλον (Enkephalon)
Cervix	Τραχηλος (Trachelos)
Cor	Καρδια (Kardia)
Glandula	’Αδην (Aden)
Ilium	’ειλεειν (eileein)
Intestinum	’Εντερον (Enteron)
Intestinum tenue	Εντερον λεπτον (Enteron lepton)
Lien	Σπλην (Splen)
Lingua	Γλοττις (Glottis)
Medulla	Μυελον (Myelon)
Musculus	Μυς (Mys)
Nervus	Νευρον (Neuron)
Oculus	’Οφθαλμος (Ophthalmos)
Omentum	’Επιπλοον (Epiploon)
Os	’Οστεον (Osteon)
Ouretera	’Ουρητηρ (Oureter)
Palatum	’Ουρανικος (Ouranikos)
Peritoneum	Περιτοναιος (Peritonaios)
Porta	Πυλορον (Pyloron)
Pulmo	Πνευμων (Pneumon)
Rectum intestinum	Προκτον (Prokton)
Ren	Νεφρος (Nephros)
Septum	Διαφραγμα (Diaphragma)
Spina	Ραχις (Rachis)
Stomachus	Στομαχος (Stomachos)
Trachea	Τραχεα (Trachea)
Urina	’Ουρον (Ouron)
Uterus	Μητρα (Metra)
Vena	Φλεψ (Phleps)
Ventriculus	Στομαχος (Stomachos)
Vertebra	Σπονδυλος (Spondylos)
Vesica	Φυσαλις (Physalis) or Φυσημα (Physema)
Viscerum	Σπλαγχνος (Splanchnos)
Vulva	Δελφυς (Delphys)

Conversely, later Greek authors such as Rufus of Ephesus (80–150 CE) and Julius Pollux (~190 CE) retained Greek terminology in their writings. Rufus's Onomasticon preserved Greek terms that are still in use today, while Pollux's Onomastikon served as a general lexicon, including medical and anatomical terms. Examples of traded Greek anatomical terms by Rufus can be seen in Table 2.

Table 2. Examples of traded Greek terms by Rufus (Toppli, 1904)

Term	Explanation by Rufus
'Ακρομιον (Akromion)	A small ossicle as described by Eudemos (nowadays Os acromiale)
'Αμνιος (Amnios)	Thin and soft skin of sheep as described by Empedokles
'Αορτη (Aorte)	Stem of all arteries, this term was created by Aristoteles
Βρογχιαι (Bronchiai)	Bronchos' extensions leading into the lungs.
Βρογχος (Bronchos)	Utilized interchangeably with Trachys arteria.
Χωριον (Chorion)	The tough outer layer surrounding the foetus in the womb contains abundant blood vessels, which develop into the umbilical cord that has two arteries and veins.
Νευρα (Neura)	Sensory nerves and fibers originating from the brain and spinal cord, as well as fibrous connections encircling and linking joints.
'Οισοφαγος (Oisophagos)	The system that allows food and beverages to descend into the abdomen; commonly referred to as Stomachos.
'Ολεκρανον (Olekranon)	The summit that holds up our reclining.
Περιοστεος (Periosteos)	Covering of additional bones
Στομαχος (Stomachos)	A synonym for esophagus.
Θυμος (Thymos)	Based on the lymphatic nodes are located at the head of the heart, near the 7th cervical vertebra and the end of the trachea, in front of the lungs.

3.1.1.4. Cultural and Mythological Influences

Anatomical terminology also reflects deep cultural and mythological roots. Terms like Atlas, Larynx, and Meninx derive from Greek mythology and literature, illustrating the interplay between language and cultural heritage. Latin terms often drew inspiration from Roman myths, such as the word Terminus, which originates from the Roman god of boundaries. In alphabetical order, examples are given in Table 3.

Oreibasios of Pergamon (c. 325–403 CE), educated in Alexandria, was one of the most prominent physicians of his time, serving as the personal doctor to Emperor Julian and overseeing the imperial library. At the emperor's request, he compiled the knowledge of earlier Greek physicians, including Galen and Rufus of Ephesus, into the monumental 72-volume *Iatrikon Synagogon* (Medical Assemblies), fragments of which survive today alongside its condensed version, the *Synopsis*, written for his son Eustathios.

Table 3. Examples of anatomical terms with mithyc origin (Karenberg, 2005)

Term	Original Meaning
Achillis tendon (Chorda Achillis)	Greek Hero (Achilles)
Arachnoidea	Theban God Amun
Atlas	Arachne, Greek Spinner
Hippocampus	Draught-animal of Poseidon's Conch
Hymen	Greek God of Wedding and Marriage; Wedding song
Iris	Messenger of Greek God Hera; Personified Rainbow
Lymphe	Clear and clean spring water
Morphe	Morpheus, one of the sons of Greek fod hypnos
Philtrum	Philtron = philtre, aphrodisiac
Pomum Adami	Hebraic "tappuac ha adam" – bump or apple on a man
Terminus	Youngest son of Greek god Saturn; identifification mark

The preservation of Greek and Latin medical texts owes much to the efforts of Arabic scholars during the Islamic Golden Age, with translators such as Rhazes, Haly Abbas, and Avicenna not only safeguarding but also expanding upon this knowledge, contributing anatomical terms like "*al-basilik*" (basilic vein) and "*al-kefal*" (cephalic vein) to modern nomenclature. The revival of human anatomy began in the 13th and 14th centuries with Mondino de' Luzzi, who authored *Anathomia*, a foundational guide for dissection that, although reliant on Galenic principles, marked the return of systematic human dissection. This progress was advanced by Jacobus Berengario da Carpi, who, through his hands-on dissection of over 100 cadavers, introduced standardised Latin terms like *Colon*, *Duodenum*, and *Jejunum*, bridging medieval and modern anatomical understanding.

The transformative work of Andreas Vesalius in the 16th century established him as the father of modern anatomy, as his *De humani corporis fabrica* (1543) replaced outdated Greek and Arabic terms with precise Latin ones, although his use of ordinal naming led to inconsistencies. His visually stunning work, illustrated by Jan Stephan van Calcar, set a new standard by merging art with science. Caspar Bauhin later refined Vesalius' nomenclature, assigning specific Latin names to anatomical structures and enhancing understanding through detailed illustrations, though some inconsistencies in naming persisted, marking a pivotal transition towards the modern study of anatomy.

3.1.2. The Birth of Nomina Anatomica

The journey towards a unified anatomical language has been long and fraught with challenges, reflecting the complex interplay of linguistic, cultural, and scientific evolution. By the seventeenth and eighteenth centuries, anatomical nomenclature had grown increasingly complicated. The addition of numerous new terms and the shift from Latin to national languages created a fragmented and inconsistent lexicon. Notable anatomists like Philip Verheyen and Josias Weitbrecht contributed to this linguistic diversity. Verheyen, a Belgian anatomist, initially published his textbook in Latin in 1699, organising terms in clear, thematic lists with illustrations. However, his 1708 German translation introduced new terms like "*Spannadern*" (tensioning vessels) for nerves, reflecting the mechanistic understanding of the time. Similarly, Weitbrecht's 1742 monograph on joint anatomy, later translated into German, not only explained existing terms but also replaced some with German equivalents, adding further inconsistency.

This chaotic state of anatomical language drew sharp criticism from Joseph Hyrtl, a nineteenth-century Austrian anatomist. Hyrtl observed that, unlike botany, where Carl Linnaeus had introduced a systematic binary nomenclature for plants, anatomy lagged in achieving linguistic standardisation. Hyrtl's works, including his *Onomatologia Anatomica* (1880), highlighted the proliferation of synonyms and eponyms, such as "*Highmors Höhle*," "*antrum Highmori*," and "*Sinus maxillaris*," which only deepened the confusion. Recognising the need for reform, Hyrtl advocated for an organised effort to resolve these inconsistencies through a dedicated committee of anatomists and linguists.

However, it was not until 1889 that decisive action was taken. The Anatomical Society formed a Nomenclature Commission, led by Albert von Koelliker, to establish a unified anatomical language. Guided by principles of assigning unique Latin names to each structure, avoiding synonyms, and maintaining linguistic accuracy, the commission developed the Basle Nomina Anatomica (BNA), which was adopted in 1895. This monumental achievement reduced the number of anatomical terms from an overwhelming 50,000 to a more manageable 5,000. Though initially conservative and not universally accepted, the BNA laid the foundation for modern anatomical nomenclature.

Reflecting on this history, it becomes clear that anatomical nomenclature is not merely a technical exercise but a cultural artefact shaped by the priorities and challenges of its time. The efforts of figures like Verheyen, Weitbrecht,

Hyrtl, and the BNA Commission underscore the need for clarity and precision in scientific communication. As junior lecturers, we can appreciate these contributions not only for their historical significance but also for their enduring relevance in shaping how we teach and understand anatomy today.

3.1.2.1. NAV Terminology

The history of anatomical nomenclature for veterinary medicine is a rich narrative of international collaboration, gradual evolution, and adaptation to the needs of diverse species. Initially, anatomical terms were inconsistent across nations, with each using its own system, often based on the discoveries of local anatomists. This led to confusion, as the same anatomical structure might have different names depending on the country.

The first major effort to standardise anatomical terms was the *Basel Nomina Anatomica* (B.N.A.) in 1895, but it was designed for humans and unsuitable for animals due to its reliance on terms based on the erect human posture. Recognising this limitation, the International Veterinary Congress attempted to create a veterinary-specific nomenclature, leading to subsequent versions like the *Nomina Anatomica Veterinaria* (N.A.V.).

Early initiatives and setbacks began in 1899 when the adoption of a veterinary nomenclature by the 7th International Veterinary Congress in Baden-Baden was pivotal but lacked widespread distribution. The American Veterinary Medical Association's 1923 publication of the N.A.V. was an important attempt but did not achieve global consensus.

Post-1950, significant strides were made when the International Association of Veterinary Anatomists (later the World Association of Veterinary Anatomists, W.A.V.A.) formed the International Committee on Veterinary Anatomical Nomenclature (I.C.V.A.N.). This committee successfully organised efforts, integrating expertise from anatomists worldwide and addressing the inadequacies of earlier systems.

From the 1960s to the 1980s, the I.C.V.A.N. established comprehensive subcommittees covering all major anatomical systems, from osteology to splanchnology and neuroanatomy. The first edition of the N.A.V. was published in 1968, followed by revisions incorporating feedback and aligning with advances in human anatomical nomenclature while maintaining veterinary-specific needs.

The evolution of veterinary anatomical nomenclature reflects the dynamic interplay between scientific rigour, educational utility, and the practical demands of veterinary medicine. It serves as a testament to the global commitment to precision and clarity in the anatomical sciences.

The N.A.V. operates on core principles designed to ensure clarity and utility. These core principles are: Uniqueness, where a single, unique term designates each anatomical structure to prevent ambiguity; Latin as Lingua Franca, with official terms being in Latin, while translations are allowed for instructional purposes; Conciseness and Simplicity, keeping terms short and straightforward; Descriptive and Instructive, where names are chosen for their educational and descriptive value, aiding memorisation and understanding; Topographical similarity, where structures related by location, such as arteries, veins, and nerves, share a root term (e.g., arteria femoralis, vena femoralis, nervus femoralis); Oppositional adjectives, where paired terms use opposing descriptors, like major/minor or superficialis/profundus; and Avoidance of eponyms, with names based on individuals being excluded in favour of terms with universal applicability.

This concise guide underpins the NAV's structure, ensuring that anatomical terms remain consistent, descriptive, and accessible across diverse veterinary contexts. The interplay between standardised principles and linguistic precision reflects the NAV's role as both a scientific tool and a linguistic bridge for veterinary anatomists worldwide. You can view the NAV via this link: https://www.wava-amav.org/wava-documents.html.

3.1.3. The Implementation of Terms for Various Types of Neurones

The classification and naming of neurones are pivotal for advancing our understanding of brain function. By integrating molecular, morphological, electrophysiological, and functional characteristics, a robust framework for neuron nomenclature can streamline communication and foster interdisciplinary collaboration in neuroscience. This chapter explores the systematic implementation of terms for various neuron types, emphasising their diversity and specialised roles in brain circuits. The classification and naming of neurones are pivotal for advancing our understanding of brain function. By integrating molecular, morphological, electrophysiological, and functional characteristics, a robust framework for neuron nomenclature can streamline communication and foster interdisciplinary collaboration in neuroscience. This chapter explores the systematic implementation of terms for various neuron types, emphasizing their diversity and specialized roles in brain circuits.

3.1.3.1. Terms for Neuron Types

A. Hippocampal Neurones

Neurones in the hippocampus are central to memory processing and spatial navigation, categorised into principal cells and interneurones. While essential, naming inconsistencies have historically hampered communication and data organisation.

B. Principal Cells

Principal cells consist of two types: Pyramidal Neurones and Granule Cells. Pyramidal Neurones, found in the CA1, CA2, and CA3 regions, are excitatory neurones that play a vital role in the hippocampal trisynaptic circuit. Meanwhile, Granule Cells, located in the dentate gyrus, contribute to the initial processing of information entering the hippocampus. Pyramidal neurones exhibit sophisticated input-output behaviours. Computational modelling using deep neural networks (DNNs) highlights their complex functional architecture, while xenotransplantation experiments in mice reveal their potential for studying human neuronal function.

C. Interneurones

Interneurones consist of two types: GABAergic Interneurones and LINCs. GABAergic Interneurones, though only 10-15% of the hippocampal neuronal population, regulate principal cells and maintain the excitation-inhibition balance. Meanwhile, LINCs are novel interneurones projecting to both local and extrahippocampal regions, influencing hippocampal oscillations.

The interneurones have key properties and functions: 1) Neurogenesis, where the dentate gyrus generates new neurones throughout life, underpinning hippocampal plasticity; 2) Molecular Profile, where unique molecular markers, such as calcium-binding proteins and receptors, define the varied functions of hippocampal neurones; 3) Electrophysiology, where distinct firing patterns and synaptic properties enable the hippocampus to encode and retrieve memories.

D. Cortical neurones

Cortical neurones are highly diverse, reflecting the complexity of the cerebral cortex. Broadly, they are classified as GABAergic interneurones and pyramidal neurones. GABAergic Interneurones, the inhibitory neurones, modulate cortical network dynamics. Advanced classification systems, such as Bayesian analysis, categorise them by molecular markers (e.g., parvalbumin, somatostatin) and their laminar-specific properties. Subtypes, such as axo-axonic cells (AACs), exhibit unique connectivity patterns, thereby further diversifying cortical networks. Meanwhile, Pyramidal Neurones, the excitatory neurones that integrate inputs from various sources, are further divided into subtypes based on their projection targets: 1) Corticocortical (CC), connecting regions within the cortex; 2) Cortico-subcortical (CS), targeting subcortical areas; 3) Corticocortical, Non-Striatal (CC-NS), a recently identified subtype with distinct connectivity and functions.

Pyramidal neurones exhibit sophisticated input-output behaviours. Computational modelling using deep neural networks (DNNs) highlights their complex functional architecture, while xenotransplantation experiments in mice reveal their potential for studying human neuronal function.

E. Type I and Type II Neurones

Neuron excitability is classified into Type I and Type II, characterised by distinct firing patterns and responses to stimuli: Type I neurones exhibit cyclic burst firing, low-threshold calcium spikes, and fast spontaneous firing rates. Found in the auditory system, they excel at encoding periodic carrier signals. In contrast, Type II neurones respond to stimuli with depolarising and hyperpolarising phases. With slower firing rates and prolonged responses, they are suited for encoding subthreshold signals in the auditory system.

F. GABAergic Interneurones

As a diverse group of inhibitory neurones, GABAergic interneurones regulate neural circuitry, maintaining the balance of excitation and inhibition. Classified by markers such as parvalbumin, somatostatin, and 5-HT3aR, they play versatile roles in shaping neural computations. Advances in single-cell analyses and interactive classification systems enhance our understanding of their properties and synaptic interactions.

GABAergic interneurones originate from progenitor cells in the subpallium, the ventral aspect of the embryonic telencephalon. Their development involves a complex migration process to the cerebral cortex, where they integrate into neural circuits. The specification and maintenance of GABAergic interneuron fate are governed by a network of transcriptional genes, which influence the interneurones' development, migration, and final distribution in the adult cerebral cortex.

G. Projection Neurones

Projection neurones are a type of neuron that sends their axons to distant targets, playing a crucial role in the connectivity and function of neural circuits. They are characterised by their long-range axonal projections, which can span various regions of the brain.

There are two types of projection neurones: corticocortical projection neurones and subcortical projection neurones. Corticocortical projection neurones are primarily found in the upper layers of the cortex and connect different regions within the cerebral cortex. They are characterised by the expression of the transcription factor Satb2, which is essential for forming axonal projections that connect the two cerebral hemispheres.

Subcortical projection neurones, located in the deep layers of the cortex, project to subcortical regions. They express transcription factors such as Sox5, Fezf2, and Ctip2, which are crucial for the specification of subcortically projecting axons.

Projection neurones arise from progenitor cells in the ventricular zones and are classified by their axonal projection patterns, which inform neural circuit organisation. High—throughput mapping techniques, such as BARseq, link transcriptomic and anatomical data to uncover principles of connectivity.

3.1.3.2. Classification Systems and Approaches

The ever-expanding field of neuroscience necessitates precise and standardised terminology for neurones to enhance communication, research, and data integration. Traditional neuron names, derived primarily from anatomical observations, often reflect subjective impressions rather than functional or molecular properties. As neuroscience progresses, it has become essential to adopt naming conventions that incorporate a comprehensive set of criteria, including gene expression, membrane properties, neurotransmitter pharmacology, and physiological characteristics. This approach ensures that neuron nomenclature aligns more closely with their functional roles and genetic profiles, facilitating effective data organisation and retrieval.

A. Gene- and Property-Based Naming

Modern neuroscience benefits from advanced tools such as RNA sequencing (RNA-seq) and optogenetics, which reveal the genetic and functional profiles of neurones. These technologies highlight the significance of molecular markers and physiological traits in neuronal classification. Efforts such as NeuronDB, NeuroLex Neuron, and Hippocampome.org have laid the groundwork for systematic neuron nomenclature, integrating morphology, connectivity, and molecular data into cohesive frameworks.

A hierarchical naming system has emerged as a promising strategy, beginning with species and brain region, followed by subregion and functional attributes. For instance, a pyramidal neuron in the primary motor cortex might be designated as "Neocortex M1 L2/3 IT pyramidal," with annotations for specific genes (e.g., Cux1) and neurotransmitters (e.g., glutamate). This methodology aligns with the evolutionary and functional organisation of neurones, providing a robust structure for interdisciplinary studies.

Despite its strengths, gene- and property-based naming faces challenges, particularly in managing the vast diversity of neuron subtypes and accounting for continuous variations in gene expression. Specialised databases and standardised abbreviations have been proposed to address these issues, ensuring both clarity and accessibility. This nomenclature has been tested across various brain regions, from simpler structures like the spinal cord to the complex neocortex. For example, cortical neurones are classified by connectivity patterns (intratelencephalic, pyramidal tract, or corticothalamic), while hippocampal neurones are categorised by regional and laminar organisation.

By integrating molecular, morphological, and functional data, this naming system supports comparative studies across species and provides a foundation for exploring neuron diversity in developmental and evolutionary contexts.

B. Neuron Phenotype Ontology (NPO)

To further address the challenges of neuron classification, the Neuron Phenotype Ontology (NPO) offers a standardised framework for defining and categorising diverse neuron types. Initiatives such as the BRAIN Initiative Cell Census Network (BICCN), Human Cell Atlas, and Blue Brain Project have generated extensive datasets on the molecular, morphological, electrophysiological, and synaptic properties of neurones. However, comparing data across laboratories using varying methodologies has proven difficult.

The NPO employs an ontology-based model that adheres to the FAIR principles (Findable, Accessible, Interoperable, Reusable). By representing neurones as collections of normalised phenotypic properties—including morphological, molecular, physiological, and connectivity characteristics—the NPO ensures compatibility across diverse classification systems. This framework bridges historical nomenclature with modern data-driven approaches, facilitating a smooth transition from traditional to contemporary classification paradigms.

The efficacy of the NPO has been demonstrated by integrating datasets from projects such as the Blue Brain Project, cortical GABAergic neuron classifications, and the Allen Institute's cell-type data. Using competency queries, researchers have shown the NPO's ability to align neuron classifications across different experimental techniques, enabling the identification of neurones with specific traits.

The NPO provides a scalable and flexible foundation for neuron classification, linking newly discovered neuron types with existing data. By bridging traditional and modern systems, it enables the scientific community to navigate the complexity of neuronal diversity, fostering deeper insights into the intricate architecture and functions of the nervous system.

3.1.3.3. Neuroanatomical Ontologies

Neuroanatomical ontologies are meticulously structured frameworks that aim to standardise the diverse terminologies and concepts used to describe the anatomy of the nervous system. These ontologies play an essential role in integrating, managing, and sharing neuroscientific data across various research domains, ensuring that complex datasets are accessible and interoperable.

A. Applications of Neuroanatomical Ontologies

One prominent ontology, the Foundational Model of Anatomy (FMA), extends its scope to include cytoarchitectural and morphological labelling schemes, such as Brodmann areas, bridging different terminologies of neuroanatomical structures. Initially designed to enrich the Unified Medical Language System (UMLS), the FMA employs disciplined modelling principles, high-level schemes, and Aristotelian definitions to establish a consistent framework. The FMA has transitioned from its original Protégé Frames language to the Web Ontology Language (OWL), addressing challenges of size and complexity through advanced reasoning tools and meticulous cleanup. Its utility extends beyond neuroscience, facilitating the modelling of pathology, physiological functions, and genotype-phenotype correlations while supporting translations into multiple languages to enhance accessibility.

In contrast, the BAMS Neuroanatomical Ontology focuses on the rat nervous system, providing detailed descriptions of neuron populations and their structural and functional relationships. It integrates structural and physiological variables within the Brain Architecture Management System (BAMS) and introduces innovative schemas to translate neuroanatomical projection reports across nomenclatures. Unlike the FMA, which excels in describing macro-level connections, BAMS incorporates "internal wiring" and functional pathways, linking individual neurones to broader brain structures. The latest version of the BAMS rat macroconnectome includes over 50,000 connectivity reports, demonstrating a rigorous and scalable approach to modelling neuroanatomical data.

Another key ontology, NeuroNames, addresses ambiguities in neuroanatomical nomenclature by providing a default vocabulary of unique structure names. It organises neuroanatomical structures hierarchically, accommodating synonyms and homonyms in multiple languages. As an indexing tool for the BrainInfo portal, NeuroNames enhances accessibility to neuroanatomical data by clarifying terminology and enabling seamless interoperability across neuroscience databases. Integrated into the FMA and UMLS, NeuroNames supports over 16,000 names linked to approximately 2,500 neuroanatomical concepts, facilitating global collaboration and data sharing across the neuroanatomical field.

B. Significance in Neuroinformatics

These ontologies—FMA, BAMS, and NeuroNames—address the inherent challenges of divergent terminologies and complex data relationships in the field of neuroscience. By offering structured, interoperable frameworks, they enable researchers to bridge the gap between traditional and modern methodologies, facilitating the exploration of brain anatomy, connectivity, and function. Their contributions to standardisation, multilingual support, and integration across platforms exemplify the critical role of ontologies in advancing neuroscience research and fostering international collaboration.

3.1.3.4. Challenges and Future Directions in Neuroanatomical Ontology Development

The development of neuroanatomical ontologies presents a unique set of challenges, primarily stemming from the need to reconcile diverse terminologies and perspectives in neuroanatomy. Historically, the absence of unified standards has hindered data interoperability across neuroscience databases, complicating collaboration and integration efforts. Ontologies like the Foundational Model of Anatomy (FMA) and NeuroNames address these issues by providing structured frameworks that harmonise varying terminologies and conceptual views. These frameworks enable consistent data annotation and facilitate seamless integration, ensuring that researchers from diverse disciplines can communicate effectively and share data without misinterpretation.

One significant barrier to progress has been the lack of terminological standards, which impedes interoperability among disparate databases. NeuroNames has emerged as a critical solution, functioning as a mediator between nomenclatures. By translating terms and codifying relationships between neuroanatomical concepts, it enhances communication and fosters data sharing across research groups with varied methodologies and focuses.

Looking forward, ontology-based models hold immense potential for advancing neuroscience research and clinical applications. These models enable the symbolic lookup, logical inference, and mathematical representation of neural systems, allowing for a deeper understanding of both normal and abnormal neural connectivity. Such capabilities are particularly valuable for practical applications, including surgical planning, where precise knowledge of neural pathways is essential, and in the development of personalised treatments for neurological disorders.

Moreover, neuroanatomical ontologies are paving the way for innovative approaches to data interaction. By supporting query-dependent visualisations, they provide intuitive, hierarchical representations of complex neuroanatomical structures. These tools not only make it easier to navigate and understand intricate datasets but also empower researchers and clinicians to interact dynamically with neuroanatomical data, enhancing both educational and investigative endeavours.

As neuroinformatics continues to evolve, the role of ontologies in addressing foundational challenges and enabling transformative applications will only grow. Their ability to integrate knowledge, support advanced modelling, and foster collaboration positions them as indispensable tools in the quest to unravel the complexities of the nervous system.

3.1.3.5. Challenges and Future Directions

The field of neuroanatomy faces significant challenges due to the diversity and complexity of the terminologies used to describe brain structures and functions. This complexity impacts data sharing, interpretation, and educational strategies. Here, we summarise the key challenges and future directions in neuroanatomical terminology based on recent research.

A. Challenges and Future Directions in Neuroanatomical Terminology

The field of neuroanatomy is inherently complex, characterised by diverse terminologies and cortical parcellation schemes used to describe brain structures and functions. This variability poses significant challenges to data sharing, interpretation, and education. The inconsistency in terminologies complicates the annotation of neuroimaging data, making it difficult to harmonise findings across studies. Traditional meta-analytic methods exacerbate this issue by focusing only on frequently occurring terms, thereby limiting their capacity to present a comprehensive understanding of brain organisation. Additionally, while magnetic resonance imaging (MRI) serves as a cornerstone tool in neuroanatomical research, its indirect measurement of biological signals introduces artefacts and confounds that can lead to misinterpretation of data.

To address these challenges, future directions in neuroanatomical terminology emphasise the integration and expansion of existing frameworks. Extending ontologies, such as the Foundational Model of Anatomy (FMA), to include diverse labelling schemes, including cytoarchitectural patterns and Broca's areas, offers a promising solution. This extension would harmonise disparate terminologies, enabling precise data annotation and fostering a deeper understanding of neuroanatomical relationships. Furthermore, ontology-based models that combine structural and functional neuroanatomical knowledge can facilitate symbolic lookups, logical inferences, and mathematical modelling, with practical applications ranging from surgical planning to personalised care for neurological disorders.

Emerging tools, such as NeuroQuery, exemplify innovative approaches to overcoming these challenges. By focusing on prediction rather than traditional inference, NeuroQuery can process text of arbitrary length, capturing rare terms and their relationships to neural correlates. This methodology provides a more comprehensive and nuanced view of brain organisation, bridging gaps across varying terminologies and datasets.

The educational domain also benefits from these advances. A multifaceted approach combining 3D visualisation tools, near-peer teaching, and mobile augmented reality offers an effective strategy for teaching neuroanatomy. These resources not only improve anatomical knowledge but also enhance long-term retention and student satisfaction, ensuring that future neuroscientists are well-equipped to navigate the complexities of the field.

As neuroanatomy continues to evolve, these innovative strategies and technologies pave the way for a more unified, accessible, and comprehensive understanding of the brain, fostering both research advancements and educational excellence.

3.2. Basic Neuroanatomy

The nervous system is a remarkable network enabling organisms to interact with and adapt to their environment. Its vast interconnected circuits support critical functions such as thought, language, sensation, learning, memory, and emotional responses. The system's adaptability is evident in its ability to respond to novel situations and recover from injuries through the plasticity of existing neural networks and the activity of neural stem cells (NSCs). These NSCs not only play a vital role during development but also aid in the regeneration and reorganisation of connections in response to environmental stimuli or damage, underscoring their importance in maintaining neural health and function.

3.2.1. Neuroanatomical Developmental Processes

The neuroanatomical development of animals, encompassing the transition from embryonic stages to adulthood, represents a complex interplay of cellular, molecular, and environmental factors. This chapter explores the developmental processes, mechanisms, and functional implications of neurogenesis, emphasising their contributions to the structure and function of the nervous system across species. By synthesising research findings, it elucidates the critical role of neurogenesis in brain development, plasticity, and adaptability.

3.2.1.1. Embryonic Neurogenesis: Foundations of Neural Development

Embryonic neurogenesis is the cornerstone of neural development, involving the proliferation, differentiation, and integration of neural progenitor cells. These processes are intricately regulated by genetic programmes and environmental cues, ensuring the precise formation of neural circuits and brain structures.

During the embryonic stage, neural progenitor cells divide symmetrically to expand their population and asymmetrically to produce a diverse range of neural cell types. This transition is orchestrated by transcription factors, epigenetic mechanisms, and signalling pathways, including BMP, Wnt/β-catenin, Notch, and Sonic Hedgehog. The interplay of these factors determines cell fate, spatial organisation, and connectivity, laying the groundwork for functional neural circuits.

Temporal dynamics characterise neurogenesis during embryonic development. For instance, granule cells in the dentate gyrus, olfactory bulb, and cerebellum initiate their development embryonically but continue postnatally. Similarly, single-cell RNA sequencing in zebrafish highlights transient neural progenitor states that evolve across developmental stages, underscoring the complexity of neurogenic waves that expand neuronal diversity.

3.2.1.2. Postnatal and Adult Neurogenesis: Continuity and Plasticity

In contrast to embryonic neurogenesis, adult neurogenesis is confined to specific neurogenic niches, primarily the subventricular zone (SVZ) and hippocampal dentate gyrus. These regions retain the capacity to generate new neurones, contributing to brain plasticity and adaptability throughout life.

Adult neurogenesis occurs in a gliogenic environment, requiring specialised mechanisms for neuronal fate specification and integration. Regulatory pathways such as BMP signalling play distinct roles, promoting quiescence to prevent neural stem cell exhaustion while maintaining neurogenic potential. Additionally, molecular regulators, including RNA-binding proteins such as HuR, mediate alternative splicing, which is critical for cognitive functions and neuronal maintenance.

The functional significance of adult neurogenesis lies in its contributions to learning, memory, and sensory processing. The integration of newborn neurones into existing circuits enhances neuronal plasticity, enabling adaptation to environmental changes. However, disruptions in these processes are linked to cognitive impairments and neurodegenerative diseases, highlighting their therapeutic potential for brain repair.

3.2.1.3. Comparative Aspects and Evolutionary Insights

The neuroanatomical development of animals varies across species, offering evolutionary insights into neural adaptation and functionality. For instance, eutherian mammals develop most brain structures embryonically, whereas marsupials exhibit significant postnatal neurogenesis. Despite these differences, lifelong neurogenesis in the subventricular zone (SVZ) and dentate gyrus is a conserved trait.

Zebrafish provide a unique model for studying neurogenesis due to their transparent embryos and regenerative capacity. Research in zebrafish reveals conserved signalling pathways and cellular mechanisms that underscore fundamental principles of neurodevelopment across vertebrates and invertebrates.

Similarly, in non-chordate deuterostomes like sea cucumbers, neuropeptides regulate nervous system development, showcasing evolutionary parallels in neurogenesis. These findings emphasise the shared molecular frameworks underlying neural development despite anatomical and functional diversity.

3.2.1.4. Molecular and Cellular Mechanisms: Regulation and Interplay

The molecular landscape of neurogenesis is a dynamic interplay of transcriptional regulators, epigenetic modifications, and environmental factors. During embryonic development, signalling pathways ensure the proliferation and differentiation of neural precursors, while in adulthood, these mechanisms adapt to maintain neurogenic potential amidst a gliogenic environment.

Critical to these processes is the cerebrospinal fluid (CSF), which acts as a neurogenic niche by providing signals that regulate the behaviour of neural precursors. This interaction persists across various life stages, influencing neurogenesis and the formation of neural circuits. Advances in single-cell transcriptomics have further unravelled the molecular signatures of neural progenitors, highlighting spatial and temporal distinctions in neurogenesis.

Oxygen availability and the HIF signalling pathway also play pivotal roles in regulating neurogenesis. Hypoxia during embryonic development activates pathways that support cell survival and differentiation, while imbalances in oxygen levels can lead to developmental disorders.

3.2.1.5. Functional Implications and Future Directions

The developmental processes of neurogenesis have profound implications for brain functionality, plasticity, and adaptability. Embryonic neurogenesis establishes the structural foundation of the brain, while postnatal and adult neurogenesis contribute to cognitive functions and sensory processing.

Understanding the regulatory mechanisms of neurogenesis offers potential therapeutic avenues for neurological disorders. Advances in neuroimaging, such as structural MRI, provide insights into the dynamics of brain maturation, enabling the characterisation of developmental variability and abnormalities.

Future research must integrate in vivo and in vitro models to elucidate the complexities of neuronal migration, differentiation, and circuit formation. Combining genetic, environmental, and computational approaches will enhance our understanding of neurodevelopment and its implications for brain health and disease.

3.2.2. Key Neuroanatomy Structure

The complexity of the nervous system is organised into distinct regions and components that enable specialised and coordinated activities.

3.2.2.1. Central Nervous System

The CNS comprises the brain and spinal cord, which are subdivided into specific regions with unique functions. The brain includes four main components: 1) the brainstem, consisting of the medulla, pons, and midbrain; 2) the cerebellum; 3) the diencephalon, which contains the thalamus and hypothalamus; and 4) the cerebral hemispheres, comprised of the cerebral cortex, basal ganglia, white matter, hippocampi, and amygdalae.

A. The Brain Stem

The brainstem, a vital structure connecting the cerebral hemispheres to the spinal cord, represents a convergence of the embryonic mesencephalon, metencephalon, and myelencephalon. Functionally, it is unparalleled among brain regions, serving as the anatomical foundation of consciousness, a regulator of autonomic processes, and the origin and target of numerous ascending and descending neural pathways. Beyond these roles, the brainstem also serves as a conduit for tracts unrelated to its intrinsic functions, houses 10 of the 12 cranial nerves and their associated nuclei, and governs critical reflexes essential for survival.

The brainstem comprises the midbrain, pons, and medulla oblongata, supporting life through its control of vital functions such as respiration and blood pressure. Central to its function is the reticular activating system (RAS), a diffuse collection of neurones along its axis that regulates consciousness, arousal, sleep-wake cycles, and muscle tone. Reflex activities such as swallowing, coughing, sneezing, and vomiting are mediated by the medulla oblongata, which also houses key cranial nerve nuclei.

The midbrain, the most minor component of the brainstem, is situated beneath the diencephalon and above the pons. It consists of the tectum and tegmentum, separated by the cerebral aqueduct (aqueduct of Sylvius). The tectum, located dorsally, includes the corpora quadrigemina, which are divided into superior and inferior colliculi. These structures coordinate head and eye movements in response to visual, auditory, and somatic stimuli, and relay auditory signals to the cerebral cortex. The ventral tegmentum contains the red nucleus, oculomotor and trochlear nerves (involved in eye movement), and the substantia nigra, which plays a key role in reward-seeking, addiction, and movement. Degeneration of the substantia nigra is a hallmark of Parkinson's disease. Additionally, the cerebral peduncles in the tegmentum connect the cerebrum to the spinal cord, facilitating motor control.

The pons, located between the midbrain and medulla oblongata, lies anterior to the cerebellum and acts as a bridge relaying information between the cerebellum and cerebrum. It also contains part of the reticular formation, which maintains wakefulness. Damage to the pons can result in conditions such as locked-in syndrome, characterised by quadriplegia and limited consciousness.

The medulla oblongata, the most inferior part of the brainstem, transitions into the spinal cord at the foramen magnum. It is critical for transmitting sensory and motor information between the spinal cord and higher brain centres. The medulla houses centres responsible for respiratory, cardiac, vasomotor, and reflexive functions. It also contains components of the reticular formation that regulate vital autonomic processes such as breathing, heart rate, and blood pressure. Given its role in life-sustaining functions, damage to the medulla often results in severe consequences, including respiratory failure or death.

B. The Cerebellum

The cerebellum, the largest structure of the hindbrain, resides in the posterior cranial fossa, situated behind the fourth ventricle, pons, and medulla oblongata. As a critical integrative hub for sensory perception and motor control, the cerebellum exhibits conserved architecture across vertebrates; however, its size, morphology, and cellular organisation vary significantly, reflecting the specific motor and sensory demands of different species. This variability underscores its adaptability in coordinating complex motor functions.

Anatomically, the cerebellum is separated from the cerebrum by the tentorium cerebelli, an extension of the dura mater. It comprises two hemispheres joined by the vermis and is further divided into three lobes—anterior, posterior, and flocculonodular—separated by two transverse fissures. The primary fissure demarcates the anterior and posterior lobes, while the posterolateral fissure divides the posterior and flocculonodular lobes. A deep horizontal fissure within the posterior lobe distinguishes its superior and inferior surfaces. This highly neuron-dense structure houses approximately 80% of the brain's neurones, organised within a uniquely convoluted cerebellar cortex.

The cerebellar cortex, a single sheet of grey matter less than 1 mm thick, folds accordion-like to form a highly compact structure. Each fold consists of an inner white matter core covered by grey matter. The grey matter is organised into three distinct layers: the outer molecular layer, the middle Purkinje cell layer, and the inner granular layer. The molecular layer contains stellate and basket cells, while the Purkinje cell layer features large, branching Golgi type I neurones. Purkinje cell axons traverse the granular layer into the white matter, where they acquire myelination and terminate in the deep cerebellar nuclei. These axons form synaptic connections with basket and stellate cells, receiving excitatory inputs from climbing and mossy fibres, which utilise aspartate and glutamate, respectively, as neurotransmitters. The climbing fibres, originating from olivocerebellar tracts, "climb" along the Purkinje dendrites. In contrast, mossy fibres represent the terminal branches of other afferent tracts, stimulating thousands of Purkinje cells through extensive branching.

Functionally, the cortex of the vermis regulates trunk movements, including those of the neck, shoulders, thorax, abdomen, and hips. Adjacent to the vermis, the intermediate zones of the cerebellar hemispheres control distal extremity movements, while the lateral hemispheres are involved in planning sequential movements and consciously assessing movement errors.

The cerebellum also features a distinct arbour vitae (Latin for "tree of life"), a branching arrangement of white matter surrounded by the grey matter of the cerebellar cortex. Embedded within the white matter are three pairs of deep cerebellar nuclei—fastigial, interposed (globose and emboliform), and dentate nuclei, the latter being the largest. Fibres from the dentate, emboliform, and globose nuclei exit the cerebellum via the superior cerebellar peduncle, whereas fibres from the fastigial nucleus utilise the inferior cerebellar peduncle. Collectively, the structural and functional organisation of the cerebellum underpins its essential role in motor coordination, error correction, and adaptive motor learning.

C. The Diencephalon, Thalamus and Hypothalamus

The diencephalon, a central region of the brain, comprises the thalamus, hypothalamus, epithalamus, and subthalamus, serving as a critical hub connecting the anterior forebrain with the rest of the central nervous system. It plays essential roles in relaying sensory and motor signals and maintaining homeostatic balance.

The thalamus, which constitutes approximately 80% of the diencephalon, is a paired grey matter structure divided into dorsal, ventral, and epithalamic regions. It serves as the primary relay station for sensory information, including visual, auditory, and somatosensory signals, directing them to the cerebral cortex. Beyond sensory relay, the thalamus influences motor systems and regulates consciousness, alertness, and sleep. Its involvement in addiction and compulsive behaviours highlights its role in linking brain regions associated with reward processing. Composed of multiple nuclei with specialised functions—such as the *ventral posterolateral* (VPL) and *ventral posteromedial* (VPM) nuclei for somatosensory pathways and the pulvinar for visual processing—the thalamus is integral to sensory integration and behavioural regulation.

Situated inferior to the thalamus, the hypothalamus is a small yet indispensable structure composed of several nuclei, including the arcuate nucleus. It governs homeostatic processes such as temperature regulation, hunger, thirst, and circadian rhythms. By linking the nervous and endocrine systems via the pituitary gland, the hypothalamus influences growth, reproduction, and stress responses. It controls autonomic functions, such as heart rate and blood pressure, through the sympathetic and parasympathetic systems. Additionally, it regulates behaviours critical for survival and reproduction, including feeding and mating, while playing a pivotal role in emotional responses. Through the synthesis and release of neurohormones, the hypothalamus orchestrates the secretion of other hormones, underpinning energy homeostasis and metabolic regulation.

The epithalamus, which includes the *habenulae* and the pineal gland, contributes to vital regulatory functions. The *habenulae* are involved in reward processing and aversive behaviours, while the pineal gland regulates circadian rhythms by secreting melatonin, thus influencing sleep-wake cycles and seasonal biological changes. Together, these components of the diencephalon underscore its central role in integrating sensory, motor, and homeostatic functions to support overall brain and body regulation.

D. The Cerebral Hemispheres

The cerebral hemispheres represent the two distinct halves of the brain, each contributing uniquely to cognitive and perceptual functions. These hemispheres are anatomically distinct and functionally specialised, facilitating diverse modes of processing information. Notable asymmetries in anatomy and neurophysiology characterise the hemispheres, a feature observed across various species from reptiles to humans. Such asymmetry underpins differential modes of attention: the left hemisphere typically supports focused, narrow attention, while the right hemisphere promotes broad, open attention. Anatomically, the left hemisphere exhibits greater grey matter relative to white matter in regions like the frontal and precentral areas, emphasising its role in localised, intra-regional processing.

Developmentally, the cerebral hemispheres grow asynchronously, with distinct growth spurts aligning with major developmental milestones. Studies utilising electroencephalographic coherence reveal that this staggered development is integral to the emergence of cognitive and perceptual abilities during early life. Functionally, the right hemisphere excels in nonverbal sound processing, visual-spatial skills, body image, and emotional perception. Damage to this hemisphere can result in cognitive and affective deficits, including hemi-inattention and prosopagnosia. Conversely, the left hemisphere dominates language-related tasks such as reading, writing, and temporal sequencing.

The hemispheres operate as independent resource systems with limited capacity, each maintaining its own resource pool that cannot be reallocated. This functional independence explains why concurrent tasks processed within the same hemisphere may interfere with one another. Structurally, the cerebral hemispheres comprise three main components: the cerebral cortex, consisting of grey matter with gyri and sulci; white matter, facilitating communication between brain regions through axons; and subcortical structures, such as the basal ganglia, thalamus, and hypothalamus, which regulate motor control, sensory relay, and homeostasis.

The cerebral cortex is divided into six lobes, each serving distinct functions: the frontal lobe governs motor control, decision-making, and problem-solving; the parietal lobe processes sensory inputs; the temporal lobe handles auditory perception and memory; the occipital lobe is dedicated to visual processing; the limbic lobe is essential for emotions and memory; and the insular lobe supports consciousness and homeostasis. Hemispheric specialisation further distinguishes the left hemisphere, known for language, analytical thinking, and fine motor skills, and the right hemisphere, associated with visuospatial abilities, attention, and emotional processing. Interhemispheric communication, facilitated by the corpus callosum, integrates cognitive and motor functions between the two sides.

Functional lateralisation within the hemispheres underscores their specialisation: language and motor functions are predominantly managed by the left hemisphere, while the right hemisphere excels in processing emotions and novel stimuli. The degree of lateralisation correlates with enhanced cognitive capabilities, such as advanced verbal and visuospatial skills, demonstrating the intricate balance and interplay between the cerebral hemispheres.

3.2.2.2. Peripheral Nervous System

The peripheral nervous system (PNS) is a critical component of the vertebrate nervous system, responsible for relaying messages between the central nervous system (CNS) and the rest of the body. It encompasses a diverse array of structures and functions, including sensory, motor, autonomic, and enteric domains.

The PNS consists of peripheral nerves, which are bundles of nerve fibres (axons) that are either myelinated or non-myelinated. These nerve fibres are organised into fasciculi, which are enclosed by three layers of connective tissue: the epineurium (surrounding the entire nerve), the perineurium (surrounding each fascicle), and the endoneurium (surrounding individual nerve fibres).

Peripheral nerves are classified into three main groups based on axonal diameter and conduction speed: A, B, and C fibres. These fibres originate from the spinal cord, where axons from the ventral roots unite with peripheral processes of dorsal root ganglia to form segmental spinal nerves.

The PNS includes ganglia, which are clusters of neuronal cell bodies. These ganglia are critical for processing and relaying information. The PNS is divided into the somatic nervous system, which controls voluntary movements, and the autonomic nervous system, which regulates involuntary functions.

There are two functional domains of the PNS: the Sensory-Motor Systems and the Autonomic Nervous System. The sensory system of the PNS includes receptors that convert mechanical and other stimuli into electrical impulses. These receptors are classified into exteroceptors (responding to external stimuli), proprioceptors (responding to movement and body position), and interoceptors (responding to internal stimuli from viscera and blood vessels). The motor system includes nerve fibres that control muscle movements, with somatic effectors being the terminals of myelinated axons of motor neurones. The Autonomic Nervous System (ANS) is further divided into the sympathetic and parasympathetic divisions. It controls various involuntary functions such as heart rate, digestion, and respiratory rate. The ANS includes preganglionic and postganglionic neurones, as well as specialised cells like adrenal chromaffin cells.

The PNS develops from various embryonic sources, including migratory neural crest cells and placodal neurogenic cells. Schwann cell precursors also play a role in the development of peripheral nerves. The PNS also exhibits significant plasticity in its neurotransmitter systems, which environmental factors can influence. This plasticity is crucial for the adaptive responses of the PNS to different physiological and pathological conditions.

3.3. Functional Neuroanatomy

Functional neuroanatomy is a highly specialised domain within neuroscience that investigates the intricate relationship between the structure of the nervous system and its functional roles. This field underpins critical advancements in areas like clinical psychology, radiology, and neurosurgery. For example, in the context of emotional processing, the medial prefrontal cortex plays a broad role across various emotional tasks. Specific emotions, such as fear and sadness, are linked to distinct neural substrates: fear is predominantly mediated by the amygdala, while sadness involves activation of the subcallosal cingulate.

Methods for inducing emotional states further highlight this interplay. Visual stimuli activate the occipital cortex and amygdala, while emotional recall or imagery engages the anterior cingulate and insula. Tasks requiring cognitive demand also recruit these regions, emphasising their integrative role in emotion and cognition. Understanding these dynamics forms the foundation for exploring cognitive and behavioural functions such as sensorimotor systems, learning, vision, attention, and language.

Functional neuroanatomy is pivotal for the design and interpretation of functional imaging studies, including fMRI and PET scans, which map brain activity related to critical functions like movement, speech, and vision. Such insights are invaluable for preoperative neurosurgical planning. Cutting-edge imaging techniques enhance this understanding, with tools like two-photon microscopy, optical coherence tomography (OCT), and Micro-Optical Sectioning Tomography (MOST) providing high-resolution, in vivo visualisation of brain structures. Advanced methods like 3D Random Access Multiphoton (3D RAMP) microscopy enable simultaneous three-dimensional recordings across multiple sites, overcoming previous temporal limitations.

Effective teaching of functional neuroanatomy demands more than rote learning; it requires practical application and critical thinking. Innovative approaches, such as interactive tools and problem-based learning, significantly enhance comprehension and student performance. In practice, tools like neuroimaging and neuropsychological assessments are indispensable. Neuroimaging methods, including fMRI and PET, reveal neural mechanisms underlying cognition and disorders, while neuropsychological assessments offer valuable insights into developmental neuroanatomy. These techniques inform interventions in clinical, school, and counselling psychology, showcasing the profound clinical and educational relevance of functional neuroanatomy.

3.4. Neuroanatomical Procedures and Tools

Neuroanatomical research employs a diverse array of methodologies and tools to unravel the complexities of the structure and function of the nervous system. These techniques range from foundational histological approaches to cutting-edge molecular, genetic, and computational innovations, each offering unique advantages in understanding neural organisation.

3.4.1. Foundational Neuroanatomical Technique

The cornerstone of neuroanatomical studies begins with tissue preparation and fixation, which preserve neural structures for downstream analysis. Techniques such as perfusion fixation for animals and the handling of unfixed fresh brain tissue are crucial for maintaining tissue integrity. Sectioning methods, including cryostat-based frozen sectioning, microtome sectioning, and vibratome sectioning, cater to different histochemical requirements. Post-sectioning enhancements, such as fluorescent Nissl staining and thionin staining, improve structural visualisation, aiding in detailed examinations of neural morphology. Autoradiography and immune histochemistry allow for the detection of specific neurochemical markers. Other techniques such as anterograde and retrograde tracers map neural pathways, those are, Phaseolus vulgaris-leucoagglutinin, biotinylated dextran amine, Fluoro-Gold. Modern ciral-based genetic tracers and barcoded viruses enable high-resolution mapping of connectivity.

3.4.2. Mapping Neural Connectivity

Neuroanatomical tract tracing has evolved significantly, bridging classical and modern approaches. Traditional tracers like Phaseolus vulgaris-leucoagglutinin (anterograde) and Fluoro-Gold (retrograde) are staples in circuit mapping. However, modern molecular-genetic techniques, such as viral vector-based delivery systems, have transformed connectivity studies. These approaches allow for unparalleled precision in tracing neural pathways, enabling insights into both local and long-range connectivity with high specificity.

3.4.3. Imaging and Computational Advancements

Imaging tools have become indispensable in neuroanatomical research and clinical applications. Structural MRI provides detailed insights into brain architecture, development, and pathology, although challenges such as artefacts and interpretative variability persist. Clinically, neuroimaging tools that utilise structural MRI and machine-learning algorithms offer diagnostic and prognostic capabilities. Despite their promise, generalisability and clinical adoption remain hurdles, emphasising the need for refined algorithms and larger datasets.

3.4.5. High-Throughput Innovations

Advances in high-throughput neuroanatomy have accelerated connectomic studies. Volume electron microscopy allows detailed neurite reconstruction, facilitated by tools like KNOSSOS software and the RESCOP procedure, which improve the efficiency and precision of reconstructing neural circuits. Additionally, the automated identification of neuroanatomical structures in PET imaging through platforms such as HERMES Brass, PMOD, and FreeSurfer is revolutionising data analysis by reducing the need for manual intervention, ensuring consistency, and saving time.

3.4.6. Data Analysis and Integration

The growing complexity of neuroanatomical data necessitates robust analytical frameworks. Tools like the natverse toolbox provide an integrative suite for analysing and visualising neuroanatomical datasets. This open-source platform supports large-scale comparisons, enabling researchers to cross-analyse data across multiple neurones and integrate diverse datasets, fostering innovation and collaboration in neuroanatomical research.

By synergising classical methods with advanced molecular, imaging, and computational techniques, the field of neuroanatomy continues to expand its horizons, offering profound insights into neural organisation and paving the way for translational applications in medicine and neuroscience.

3.5. Molecular Neuroanatomy

Molecular neuroanatomy represents a dynamic intersection of molecular biology and neuroanatomy, enabling a nuanced understanding of the architecture and functionality of the nervous system at the molecular level. This field, empowered by rapid technological advancements, has transformed the landscape of neuroscience, providing unprecedented insights into neuronal structure, activity, and connectivity.

3.5.1. Technological Milestones in Molecular Neuroanatomy

The advent of cutting-edge tools has redefined molecular neuroanatomical research. Brain gene expression atlases now offer comprehensive, high-resolution maps of gene activity across diverse brain regions, revealing molecular signatures unique to specific neuronal types and anatomical domains. Genetically encoded proteins, such as calcium indicators and optogenetic actuators, have become indispensable for monitoring and modulating neuronal activity with remarkable precision, enabling real-time exploration of complex neural dynamics. Imaging innovations like spatial transcriptomics have revolutionised the mapping of brain circuits by integrating molecular data with spatial resolution, yielding intricate molecular blueprints that elucidate tissue organisation and connectivity.

3.5.2. Spatial Transcriptomics: A Paradigm Shift

Spatial transcriptomics stands at the forefront of molecular neuroanatomy, offering transformative insights into brain architecture. By enabling unsupervised and unbiased gene expression profiling, this technique categorises brain regions at a molecular level, transcending traditional anatomical boundaries. It creates reference frameworks that integrate cellular composition, connectivity, and functional activity, paving the way for novel and more detailed brain maps. This molecular classification not only deepens our understanding of the nervous system but also bridges gaps in linking structure to function.

3.5.3. Viral Tracing Techniques: Unravelling Neural Pathways

The application of neurotropic viruses in connectivity studies has revolutionised our ability to trace neural circuits. These viruses facilitate genetic material transfer across synaptically connected neurones, uncovering multi-synaptic pathways within and between brain regions. Advances in recombinant viral engineering now allow precise targeting of genetically defined neuronal populations, enhancing specificity and accuracy in connectivity mapping. These methods provide critical insights into the fundamental organisation of neural networks and their roles in behaviour and cognition.

3.5.4. Interdisciplinary Innovations: Expanding Horizons

The integration of diverse methodologies has significantly broadened the scope of molecular neuroanatomy. Techniques such as dye injection and immunocytochemistry combine anatomical, electrophysiological, and immunological approaches, offering comprehensive views of neuronal morphology, function, and chemistry. Computational advancements further enrich this field, as algorithms merge genetic, transcriptomic, and connectivity data, enabling high-throughput functional analysis. These approaches predict functional neuroanatomical maps, identify behaviourally or clinically relevant neural networks, and facilitate targeted therapeutic strategies for neurological and psychiatric conditions.

Molecular neuroanatomy exemplifies the power of interdisciplinary science, linking molecular detail with system-wide insights. As tools and techniques continue to evolve, this field holds promise for uncovering the intricate mechanisms underlying neural function and dysfunction, driving innovations in neuroscience research and clinical practice.

Conclusion

A transparent and standardised neuroanatomical language is essential for fostering accurate communication, reproducible research, and effective clinical practice across biomedical disciplines. This chapter has traced the historical progression of neuroanatomical terminology, beginning with classical Latin and Greek origins and evolving toward modern ontology-based systems that prioritise consistency, interoperability, and digital integration.

The refinement of anatomical terms has not only improved clarity. Still, it has also enabled the alignment of veterinary and human anatomical nomenclature, facilitating cross-species comparisons, translational research, and collaborative diagnostics. Classification systems and ontologies serve as critical scaffolds that underpin the organisation of neuroanatomical data, supporting the development of atlases, computational models, and educational platforms.

Furthermore, the integration of these frameworks with emerging technologies—such as high-resolution imaging, transcriptomics, and informatics has expanded the utility of neuroanatomical knowledge beyond traditional descriptive roles. They now play a pivotal part in linking structure to function, aiding in disease modelling, and guiding precision interventions in both experimental and clinical settings.

In summary, neuroanatomical standardisation is not a static achievement but a dynamic foundation that evolves alongside scientific progress. By bridging historical context, structural classification, and technological innovation, it continues to support the advancement of neuroscience across veterinary and human medicine, shaping a common language for the exploration and treatment of the nervous system.

Chapter 4

The Brain's Radar – Unveiling the Sensory System

The sensory system serves as the gateway between the external environment and the internal processes of the nervous system, facilitating the detection, transmission, and interpretation of sensory stimuli. This intricate network integrates diverse sensory modalities (vision, hearing, olfaction, taste, somatosensation, proprioception, and interoception) into the central nervous system (CNS). Through this integration, sensory systems influence not only perception but also higher cognitive functions and behaviour.

Advances in neuroimaging and electrophysiological methods have significantly enhanced the understanding of sensory systems. These technologies have unveiled the hierarchical nature of sensory processing, from peripheral sensory detection to cortical integration. Furthermore, the study of sensory systems has revealed their role in maintaining homeostasis by regulating autonomic functions, such as heart rate and respiratory control, in response to external stimuli. The dual role of sensory systems—encompassing perception and homeostatic regulation—highlights their complexity and importance in overall nervous system functionality.

4.1. Components and Pathways in Sensory Systems

4.1.1. Peripheral Receptors and Connection Pathways

Peripheral sensory receptors serve as the entry point for sensory systems, designed to detect a wide range of stimuli. Mechanoreceptors, for example, can differentiate between light touch and deep pressure, whereas nociceptors respond to potentially damaging stimuli, triggering protective behaviours. These receptors are distributed across the body in specialized organs, such as the skin, retina, and cochlea, tailored to their respective sensory modalities. Their ability to transduce physical or chemical signals into electrical activity underscores their fundamental role in sensory systems.

The pathways connecting peripheral receptors to the central nervous system (CNS) are equally specialised. Axons from these receptors form afferent nerves that transmit signals to the spinal cord or brainstem. For instance, in the somatosensory system, signals travel via the dorsal columns to reach the thalamus and then the somatosensory cortex. The precision of this transmission ensures that sensory inputs are delivered to appropriate cortical areas for detailed processing. Disruptions in these pathways, as seen in conditions like multiple sclerosis, can lead to sensory deficits, emphasising their critical role in normal sensation and perception.

4.1.2. Cortical Processing Centres

Cortical processing centres in the brain are responsible for interpreting sensory inputs and generating meaningful perceptions. These centres are organised topographically, with different regions dedicated to specific sensory modalities. The visual cortex, for instance, processes spatial and temporal information about the environment, while the auditory cortex decodes sound frequencies and patterns. The olfactory cortex, on the other hand, integrates chemical signals from the environment to generate complex odour perceptions, often linked to memories and emotions.

Importantly, these processing centres do not work in isolation. Cross-modal integration between different cortical areas allows for a richer sensory experience. For example, visual and auditory information can combine to enhance speech recognition in noisy environments. This interplay is facilitated by a network of interconnections within the brain, enabling higher-order functions such as attention, learning, and decision-making.

4.2. Functional Characterisation and Modelling of Sensory Systems

4.2.1. Quantitative Models and Neural Representations

Quantitative models of sensory systems provide a framework for understanding how neurons encode information about the environment. These models are constructed using data from behavioural experiments, electrophysiological recordings, and imaging studies. For example, researchers have used these models to predict how retinal neurons respond to varying light intensities or how auditory neurons encode different sound frequencies. By capturing these dynamics mathematically, scientists can simulate sensory processing under normal and pathological conditions.

Neural representations derived from these models reveal how sensory inputs are transformed into perceptions. In the visual system, for instance, receptive fields in the primary visual cortex respond to specific orientations or motion directions. Similar transformations occur in other sensory systems, such as the auditory cortex, where neurons are tuned to particular sound frequencies. These representations are not static; they are shaped by experience, learning, and environmental changes, reflecting the adaptive nature of sensory systems.

4.2.2. Neuronal Circuits and Sensory Hierarchies

Neuronal circuits in sensory systems are organised hierarchically, enabling progressively complex processing. In the somatosensory system, for example, primary afferents detect basic stimuli, such as pressure or vibration. At the same time, higher-order neurons integrate these inputs to generate more complex perceptions, including texture or shape recognition. This hierarchical organisation allows for efficient processing, with each level extracting relevant features from the sensory input.

At the highest levels, these circuits contribute to perceptual decision-making and behaviour. The integration of sensory inputs with motor and cognitive systems enables adaptive responses to environmental changes. Understanding these circuits has implications for developing interventions for sensory disorders, such as using brain-machine interfaces to restore lost sensory functions in patients with spinal cord injuries.

4.3. Plasticity and Adaptation in Sensory Systems

4.3.1. Receptive Field Dynamics and Synaptic Modulation

Receptive fields, which represent the specific area or modality to which a neuron responds, are not fixed but exhibit remarkable plasticity. For example, in the visual system, neurons in the primary visual cortex can adjust their receptive fields in response to prolonged exposure to specific stimuli. This dynamic property allows sensory systems to remain responsive to changes in the environment, such as altered lighting conditions or new sensory inputs.

Synaptic plasticity plays a crucial role in this adaptability. Changes in synaptic strength, driven by factors such as long-term potentiation (LTP) and long-term depression (LTD), enable the sensory cortices to refine their responses in response to experience. Neuromodulators such as dopamine and acetylcholine further influence this process, enhancing plasticity during learning or attention-demanding tasks. These mechanisms ensure that sensory systems are not only reactive but also capable of predictive and anticipatory adjustments.

4.3.2. Integration and Cross-Modal Adaptation

Sensory integration, where information from multiple modalities is combined, enhances perception and behaviour. For example, visual and proprioceptive inputs are integrated to guide movements, such as reaching for an object. Cross-modal plasticity, observed in individuals with sensory deficits, highlights the brain's ability to adapt by enhancing the functionality of other modalities. For instance, individuals who are blind often exhibit heightened auditory or tactile sensitivity, demonstrating the compensatory mechanisms within the central nervous system.

This adaptability extends to the sensorimotor network, where sensory feedback is crucial for fine-tuning motor actions. Studies on athletes and musicians have demonstrated that repetitive practice enhances sensory-motor integration, resulting in improved performance. These findings underscore the importance of sensory plasticity in skill acquisition and rehabilitation.

4.4. Challenges and Advances in Sensory Neuroscience

4.4.1. Technological Progress in Neural Recording

The advent of advanced neural recording techniques has revolutionised sensory neuroscience. Methods such as two-photon microscopy and optogenetics allow for high-resolution imaging and manipulation of neural circuits in living organisms. These technologies have been instrumental in mapping sensory pathways and understanding their dynamics at the cellular and network levels. For instance, calcium imaging has revealed the activity of entire neuronal populations during sensory processing, offering unprecedented insights into how sensory information is encoded and transmitted.

Beyond recording, tools such as spatial transcriptomics and single-cell RNA sequencing offer molecular-level insights into sensory neurons. These methods have uncovered distinct subpopulations of sensory neurons with specialised functions, paving the way for targeted therapies in sensory disorders. The integration of these techniques with computational models has further accelerated progress in the field, enabling the prediction of neural responses under various conditions.

4.4.2. Sensory System Plasticity and Rehabilitation

The understanding of sensory plasticity has led to the development of innovative rehabilitation strategies for individuals with sensory deficits. Techniques such as sensory retraining, which involves repetitive exposure to stimuli, have shown promise in restoring function after injuries like stroke. Neuroprosthetics and brain-machine interfaces have also emerged as transformative tools, enabling patients to regain lost sensory or motor functions through the use of artificial devices.

Additionally, research on adaptive plasticity has implications for treating chronic pain and sensory hypersensitivity. By modulating neural circuits involved in sensory processing, it may be possible to alleviate these conditions and improve quality of life. These advancements highlight the potential of sensory neuroscience to address pressing clinical challenges and enhance human health.

4.5. Emerging Areas of Research in Sensory Neuroscience

4.5.1. Temporal Dynamics and Predictive Coding

The study of temporal dynamics and predictive coding has unveiled how the brain anticipates sensory events. Predictive coding frameworks suggest that the brain generates internal models to forecast incoming stimuli, minimising the cognitive effort required for processing predictable inputs. This mechanism is evident in phenomena such as auditory beat perception or visual tracking of moving objects.

Research on temporal coding in neuronal activity has also highlighted the importance of timing in sensory processing. For instance, the precise timing of spikes in auditory neurones underpins speech recognition and music perception. These findings have implications for the development of auditory prosthetics, such as cochlear implants, that mimic natural timing patterns to enhance sound quality and improve speech comprehension.

4.5.2. The Role of Sensory Systems in Emotional and Cognitive Processes

Sensory systems are deeply intertwined with emotional and cognitive functions. The olfactory system, with its direct connections to the limbic system, plays a crucial role in forming memories and regulating emotional responses. Similarly, the auditory system's involvement in speech and language underscores its significance in social communication.

Recent research has explored how sensory deficits impact mental health, revealing links between sensory processing disorders and conditions such as anxiety, autism, and depression. Understanding these connections may lead to therapies that address both sensory and emotional challenges, fostering holistic approaches to mental health care.

Conclusion

The sensory system acts as the brain's gateway to the external world, translating environmental stimuli, such as light, sound, pressure, and chemical signals, into meaningful neural messages that inform perception, guide behaviour, and maintain physiological homeostasis. This chapter has examined the anatomy and function of key sensory modalities, including vision, audition, somatosensation, olfaction, and gustation, highlighting the intricate neural pathways that carry information from peripheral receptors to specialised cortical processing centres.

Beyond basic sensory transduction, the chapter has also underscored the role of neural plasticity in modulating sensory processing. Experience-dependent changes in synaptic strength, cortical reorganisation following injury, and the influence of learning and attention demonstrate how dynamic and adaptable the sensory system truly is. These mechanisms not only optimise sensory perception in response to environmental demands but also underlie critical aspects of development, emotional regulation, and cognitive function.

Significantly, advances in imaging, electrophysiology, and computational modelling have significantly deepened our understanding of sensory circuits and their dysfunctions. Such progress informs clinical approaches to sensory impairments, from cochlear implants and retinal prostheses to neuromodulation therapies for chronic pain and sensory processing disorders. As research continues to bridge basic neuroscience and translational medicine, the sensory system remains a central focus for both understanding brain function and developing strategies to restore or enhance it.

In essence, the sensory system does more than detect the world, but it also shapes our experience of it. This study not only reveals the fundamental principles of neural coding and perception but also offers powerful pathways toward improving quality of life across the lifespan.

Chapter 5

Neuroplasticity – A Comprehensive Exploration of Its Mechanisms, Influences, and Applications

Neuroplasticity refers to the brain's remarkable capacity to reorganise and adapt by forging new neural pathways. It underpins learning, memory formation, and recovery from neurological injuries. This dynamic ability enables the brain to adjust in response to experience, environmental input, and damage, forming a foundational principle in cognitive development and survival. Neuroplasticity manifests through synaptic, structural, and functional transformations, each contributing distinctively to neural adaptability.

A fundamental component, synaptic plasticity, involves the modulation of synaptic strength. Processes such as long-term potentiation (LTP) and long-term depression (LTD) exemplify this mechanism. LTP enhances synaptic transmission following high-frequency stimulation, a key element in encoding memory and learning. LTD, conversely, reduces synaptic strength, facilitating synaptic pruning and memory refinement. These processes form the molecular architecture of neuroplasticity, highlighting the brain's equilibrium between synaptic reinforcement and elimination.

Structural plasticity involves modifications to neural architecture, including dendritic spine proliferation, neurogenesis, and synaptogenesis. These changes are not restricted to early development but persist throughout life, albeit more gradually in adulthood. Such alterations underscore the brain's enduring capacity for adaptation, learning, and functional recovery.

5.1. Mechanisms Underpinning Neuroplasticity

5.1.1. Cellular and Molecular Processes

Neuroplasticity reflects the nervous system's ability to adapt both structurally and functionally to intrinsic and extrinsic cues. At the cellular and molecular level, this includes synaptic reorganisation, dendritic spine modulation, adult neurogenesis, and gene expression that is dependent on neural activity. Schwann cells and oligodendrocytes play significant roles in remyelination and structural adaptation within both peripheral and central nervous systems. Neural stem cells (NSCs) and precursor cells (NPCs) facilitate ongoing neurogenesis via regulated transcriptional cascades, influenced by key signalling pathways such as BDNF, IGF, Wnt, and Notch.

Plasticity of dendritic spines, particularly in regions such as the hippocampus and cortex, enables rapid synaptic alterations that support learning. Similarly, LTP and LTD contribute to synaptic modulation that underpins the broader functional capabilities of the nervous system.

5.1.2. Types of Plasticity

Neuroplasticity is broadly classified into functional and structural forms. Functional plasticity entails changes in the efficiency and connectivity of neural networks, exemplified by mechanisms such as LTP and LTD. These contribute to the dynamic reorganisation of synaptic strength and functional circuit properties. Structural plasticity refers to anatomical adaptations including synaptogenesis, dendritic branching, and remodelling of myelin.

Adult neurogenesis, particularly in the subventricular zone and hippocampal dentate gyrus, demonstrates the continuity of plasticity throughout life. Moreover, this capacity extends beyond grey matter, as shown by environmentally induced alterations in myelin.

5.2. Factors Influencing Neuroplasticity

Both intrinsic and extrinsic influences modulate neuroplasticity. Age is a critical factor; heightened plasticity is observed during early development but gradually declines. Nonetheless, contemporary evidence confirms the persistence of plasticity throughout the lifespan, supporting rehabilitation and cognitive enhancement in later life.

5.2.1. Developmental Windows

Critical and sensitive periods during early neurodevelopment represent phases of heightened malleability. For instance, synaptogenesis during infancy enables the rapid establishment of sensory and cognitive networks. Perturbations during these windows (genetic, nutritional, or environmental) can result in enduring developmental disorders, as observed in conditions such as autism spectrum disorder and attention-deficit hyperactivity disorder.

Although these sensitive periods eventually close, plasticity remains, albeit under stricter regulation by activity-dependent signalling and inhibitory processes. This shift delineates the transformation from a developmentally open to a more selectively adaptable brain.

5.2.2. Lifestyle and Environment

Environmental enrichment, physical activity, and sensory input promote adaptive plastic changes. Aerobic exercise, in particular, elevates hippocampal BDNF, thereby enhancing neurogenesis and cognitive flexibility. Sleep is another vital component, essential for memory consolidation and synaptic recalibration. Its neglect is detrimental to neuroplastic efficiency.

Emerging studies also point to the influence of the gut–brain axis. Microbiota-derived metabolites affect neurotrophic signalling and neuroinflammatory states, further highlighting the multifaceted nature of plasticity regulation.

5.2.3. Stress and Disease

Chronic stress and neurodegenerative conditions significantly compromise neuroplastic capacity. Prolonged exposure to glucocorticoids, oxidative stress, and inflammatory processes reduces synaptic resilience and impedes neurogenesis. This leads to neuronal atrophy and synaptic degradation, particularly in the hippocampus and prefrontal cortex.

Nevertheless, several interventions have shown promise in restoring plasticity. Pharmacological agents, cognitive stimulation, and neuromodulation techniques offer partial recovery of function. Disorders such as major depressive disorder, Alzheimer's disease, and stroke are increasingly understood as disorders of disrupted plasticity, thereby underscoring the therapeutic relevance of restoring neural adaptability.

5.3. Applications and Therapeutic Implications

An understanding of neuroplasticity has revolutionised rehabilitation and therapeutic approaches. In neurology and psychiatry, plasticity is the basis for interventions aimed at re-establishing functional circuits. Recovery following cerebrovascular insult, for instance, depends on reorganisation of surviving neural networks, facilitated by targeted physiotherapy and neuromodulatory techniques.

In neuropsychiatric care, the mechanisms of selective serotonin reuptake inhibitors (SSRIs) and ketamine are believed to involve the enhancement of synaptic plasticity and increased expression of BDNF. Techniques such as transcranial magnetic stimulation (TMS) and transcranial direct current stimulation (tDCS) exploit these mechanisms to improve clinical outcomes.

In the context of mental illness, disrupted plasticity is implicated in anxiety, depression, and post-traumatic stress disorder. Treatments designed to restore neural adaptability, through behavioural therapy, pharmacology, and lifestyle change, are proving to be effective avenues for intervention.

Within neurodevelopmental and cognitive impairment contexts, plasticity-driven therapies such as environmental enrichment, brain–computer interface integration, and multisensory stimulation are utilised to recalibrate pathological circuits. Advances in neuroimaging, such as diffusion tensor imaging (DTI) and functional MRI (fMRI), now permit the observation and modulation of plastic changes in real time. Comparative studies using animal models complement this by elucidating molecular mechanisms with translational relevance.

Finally, regenerative medicine is increasingly guided by principles of plasticity. Stem cell-based interventions seek to replace lost neuronal populations and reconstruct neural architecture. When paired with tailored rehabilitation programmes, these strategies hold significant promise for maximising intrinsic repair capabilities.

5.4. Challenges and Future Directions

Despite its transformative potential, neuroplasticity research faces notable obstacles. The complexity of neural systems and individual differences in response to intervention call for a personalised and integrative approach. Bridging insights from molecular biology, computational modelling, behavioural science, and clinical practice remains a significant but necessary challenge.

Future inquiry should prioritise the development of interventions that harness neuroplasticity across diverse contexts. This includes the refinement of pharmacological agents targeting specific molecular pathways, and the design of customised physical and cognitive training regimens. Furthermore, the application of artificial intelligence and machine learning may offer predictive models for plasticity-driven outcomes, enhancing precision and efficacy in therapeutic deployment.

Conclusion

Neuroplasticity stands as one of the most compelling principles of contemporary neuroscience. It encapsulates the brain's intrinsic capacity to adapt, reorganise, and repair itself in response to experience, environmental change, and injury. From the molecular intricacies of synaptic modulation to the broader architecture of circuit reconfiguration, plasticity underscores the dynamic nature of neural systems.

Throughout life, neuroplastic mechanisms support not only learning and memory but also recovery and resilience in the face of pathology. Advances in cellular biology, neuroimaging, and behavioural science have deepened our understanding of the factors that influence this process (critical developmental windows to lifestyle, stress, and disease).

As science moves forward, the therapeutic implications of neuroplasticity continue to grow. It is now central to rehabilitation strategies, psychiatric treatment, and regenerative medicine. Yet, with this promise comes complexity. Personal variability, disease burden, and the intricate organisation of the nervous system pose significant challenges to generalisation and clinical application.

To harness neuroplasticity effectively, research must remain integrative, ethically grounded, and responsive to emerging technologies. In doing so, we not only advance scientific knowledge but also offer renewed hope for recovery, adaptation, and enhancement across the lifespan.

Chapter 6

Behavioural Neuroscience – A Comprehensive Exploration

Behavioural neuroscience is a complex and continually evolving discipline that investigates the neural substrates underlying normal and pathological behaviours in animals. By drawing upon neurobiology, psychology, pharmacology, and computational sciences, this field aims to elucidate how the brain and nervous system shape behavioural output. It forms a crucial link between molecular mechanisms and clinical outcomes, thereby playing an instrumental role in the development and assessment of therapeutic interventions for neuropsychiatric disorders. This chapter presents a detailed examination of behavioural neuroscience through its principal components, each underpinned by current scientific perspectives.

6.1. Neural Mechanisms and Behaviour

Behavioural neuroscience examines the intricate neural processes responsible for generating behaviour, emphasising how brain activity governs responses ranging from basic reflexes to higher cognitive functions. Behaviour does not emerge from isolated regions but from integrated networks involving sensory processing, motor control, memory, and emotion. The central nervous system, with its extensive synaptic interconnections, orchestrates the interpretation of sensory stimuli and the coordination of appropriate responses.

Key regions include the prefrontal cortex, which underpins decision-making and executive function, and the limbic system, which regulates emotions and social conduct. The striatum, a subcortical nucleus, is essential for reward learning and habit formation, and is heavily implicated in understanding addiction and compulsive behaviours.

The advent of advanced techniques such as in vivo two-photon microscopy, optogenetics, and calcium imaging has enabled real-time observation and manipulation of neuronal circuits. These technologies have significantly advanced our understanding of brain–behaviour relationships.

For instance, optogenetic interventions allow researchers to activate or silence neuronal populations selectively, providing causal evidence for the neural regulation of specific behaviours such as aggression and social bonding.

Genetic studies complement these findings by highlighting inherited traits that influence neural and behavioural phenotypes. Circadian rhythms, anxiety tendencies, and social preferences are among behaviours shaped by genetic regulation, offering insights into developmental and psychiatric conditions.

6.2. Animal Behaviour Paradigms

Experimental paradigms involving animal behaviour are foundational to behavioural neuroscience. These models facilitate the translation of basic neural mechanisms into clinical relevance. Paradigms range from simple assessments of locomotion to complex tasks examining cognition, emotion, and social dynamics. For example, the Morris water maze is widely employed to investigate spatial learning and memory, whereas the open field test evaluates anxiety and exploratory tendencies.

Tests such as the elevated plus maze or social defeat paradigm provide robust models for studying affective and social behaviours, offering valuable proxies for human psychiatric symptoms. Increasingly, researchers seek to enhance the ecological validity of these models by simulating more naturalistic conditions. This shift ensures that observed behaviours closely reflect those exhibited in the wild, improving translatability.

Modern behavioural assays now incorporate technologies such as virtual reality and automated motion tracking to record behaviour in a minimally invasive manner. These tools, along with computational modelling, allow for sophisticated behavioural analysis and prediction based on neural activity patterns.

6.3. Translational Research and Clinical Applications

Translational research in behavioural neuroscience seeks to bridge the gap between experimental findings and clinical practice. Animal models are instrumental in exploring the pathophysiology of neuropsychiatric disorders such as depression, schizophrenia, and anxiety. For instance, chronic social defeat in rodents is used to mimic depressive phenotypes and test antidepressant efficacy.

These models have guided the identification of key neurotransmitter systems (serotonin, dopamine, and noradrenaline) that regulate mood and affective states. Genetic engineering techniques, including CRISPR-Cas9, now enable the creation of animal models carrying human-specific mutations, thus increasing the translational value of preclinical studies.

Brain imaging methods, such as functional MRI and PET, complement behavioural assessments by revealing activity patterns linked to specific behaviours. This multimodal approach aids in mapping brain circuits implicated in psychiatric conditions and enhances the understanding of pharmacological or behavioural interventions at a systems level.

6.4. Methodological Approaches and Technological Innovations

Behavioural neuroscience relies on a variety of methodological approaches, ranging from traditional techniques such as electrophysiology and lesion studies to modern technologies like optogenetics and CRISPR-Cas9. The choice of approach often depends on the research question at hand. For instance, somatic interventions such as pharmacological manipulations allow for the study of the effects of neurotransmitters or drugs on behaviour, while behavioural interventions, including training and conditioning protocols, reveal how modifications in behaviour influence neural activity.

The advent of new technologies has expanded the capacity of researchers to observe and influence behaviour with unprecedented precision. For example, miniaturised neural recording devices and wireless brain (computer interface) enable the continuous monitoring of brain activity while animals move freely and interact with their environment. These developments have significantly improved the ecological validity of behavioural studies. In addition, computational neuroscience has given rise to sophisticated models that simulate neuronal networks, providing insights into how populations of neurones communicate and coordinate to produce behaviour. These computational models not only enable the testing of hypotheses that may be impractical or impossible in live animals, but also contribute to a deeper understanding of complex cognitive and motor processes.

Machine learning algorithms have become increasingly valuable for analysing large-scale data sets, such as those generated by video tracking systems that record behaviour over extended periods. These algorithms can identify patterns and classify behaviours in a way that manual coding cannot achieve, resulting in faster and more reliable data analysis. The integration of machine learning with real-time monitoring and intervention allows researchers to investigate how behaviours evolve, providing a powerful tool for studying the impact of environmental or genetic factors on behaviour.

6.5. Challenges in Behavioural Neurosciences

While behavioural neuroscience has made substantial strides, several challenges persist. A significant issue is the reductionist approach, which, although helpful in isolating variables, may obscure the integrative nature of neural function. There is growing recognition of the need for systems-level analysis that considers the interplay among multiple brain areas.

Ethical considerations are also paramount. Although animal models provide invaluable insights, researchers are obligated to minimise distress and adhere to the principles of Replacement, Reduction, and Refinement. These principles promote ethical conduct while preserving scientific integrity.

Future directions must involve the integration of genetic, epigenetic, environmental, and behavioural data. The development of refined tools and interdisciplinary approaches (neuroscience with engineering, psychology, and computational science) will be critical for advancing our understanding of brain–behaviour relationships and improving interventions for neuropsychiatric conditions.

Conclusion

Behavioural neuroscience serves as a vital link between the structure and function of the brain and the behaviours it governs. By combining neurobiological techniques with behavioural analysis, the field provides a deeper understanding of how neural circuits generate, modulate, and sustain complex behaviours. Innovations such as optogenetics, machine learning, and in vivo imaging have transformed the ability to observe brain–behaviour relationships with unprecedented precision.

As behavioural neuroscience continues to evolve, its translational relevance grows stronger. From refining diagnostic tools to developing personalised interventions for neuropsychiatric conditions, the field holds the potential to improve both scientific knowledge and clinical outcomes significantly. A sustained commitment to methodological rigour, ethical responsibility, and interdisciplinary collaboration will be essential to realise its promise fully.

Chapter 7

Autonomic Nervous System – Anatomy, Function, and Complexity

The autonomic nervous system (ANS) constitutes a vital division of the vertebrate nervous system, orchestrating involuntary physiological functions essential to survival. These include cardiac rhythm, respiratory dynamics, digestion, pupillary reflexes, urination, and sexual function. It comprises three major subdivisions: the sympathetic, parasympathetic, and enteric systems. Collectively, these divisions ensure homeostatic regulation and mediate adaptive responses to internal and external stimuli. This chapter explores the anatomical organisation, functional principles, and developmental aspects of the ANS in animals, integrating contemporary evidence and expert analyses to illustrate the system's sophistication and regulatory depth.

7.1. Anatomy and Physiology of the Autonomic Nervous System

The sympathetic, parasympathetic, and enteric branches of the ANS perform distinct yet interconnected roles. The sympathetic system originates from the thoracolumbar spinal segments and mediates the classic 'fight or flight' response, characterised by elevated heart rate, bronchodilation, and mobilisation of energy reserves. The parasympathetic division arises from craniosacral regions and facilitates the 'rest and digest' state, slowing cardiac activity and promoting digestion. The enteric nervous system, embedded within the gastrointestinal tract and often referred to as the 'second brain,' operates independently of central control, though both sympathetic and parasympathetic inputs modulate it. It provides localised regulation of gut motility and secretory activity.

Neurotransmitter profiles further differentiate these systems. Sympathetic postganglionic fibres predominantly release noradrenaline, whereas parasympathetic terminals utilise acetylcholine. The enteric system employs an array of neurotransmitters including serotonin and nitric oxide, reflecting its diverse functionality in gastrointestinal regulation. Advances in neural tracing and imaging have enhanced our understanding of these systems' structural and functional integration. Sympathetic networks exhibit complex, hierarchical organisation with long postganglionic projections, while parasympathetic pathways are typically more direct. The enteric system comprises intricate plexuses (myenteric and submucosal) that enable a high degree of autonomy in digestive control.

7.2. Central Processing and Regulation of Autonomic Function

Autonomic control is centrally orchestrated through the central autonomic network (CAN), encompassing brain regions such as the amygdala, insula, and midcingulate cortex. Functional imaging in both humans and animal models reveals task-specific activations: the insula, for example, processes interoceptive signals across diverse contexts, whereas the midcingulate cortex responds during stress and cognitive demand. Sympathetic activity is modulated by executive and salience networks, while parasympathetic tone aligns more closely with the default mode network.

Animal studies employing electrophysiology and neurochemical profiling have uncovered core pattern generators in the medulla and hypothalamus. Notably, the paraventricular nucleus of the hypothalamus integrates visceral afferent input to modulate both branches of the ANS, reflecting its central role in adaptive control. The Polyvagal Theory, an evolutionary model, posits that the mammalian vagus nerve, particularly its myelinated branch, supports both physiological regulation and social engagement. This perspective links autonomic function with affective and behavioural domains, offering a neurobiological framework for understanding complex emotional processes.

7.3. Role in Homeostasis and Adaptive Responses

The ANS is indispensable for maintaining internal stability. It continuously adjusts cardiovascular, respiratory, gastrointestinal, and thermoregulatory functions in response to changing physiological and environmental conditions. For instance, baroreceptor reflexes finely balance sympathetic and parasympathetic influences to maintain arterial pressure. Similarly, respiratory control centres in the brainstem modulate breathing patterns in accordance with metabolic demands.

The system's adaptability is particularly evident under stress. Acute stress triggers a sympathetic response geared toward immediate survival, while parasympathetic activity facilitates post-stress recovery. However, prolonged or repeated stress exposure can disrupt autonomic balance, contributing to pathophysiological states including hypertension, cardiac arrhythmias, and metabolic dysfunction. Animal models have been invaluable in delineating these dynamics and identifying potential therapeutic targets for restoring autonomic equilibrium.

7.4. Development, Vulnerability, and Disorders

The development of the ANS is a finely regulated process beginning in early embryogenesis. Neural crest derivatives and placodal cells differentiate into autonomic neurons under the influence of transcriptional regulators and environmental cues. This process is highly susceptible to perturbation. Prenatal exposure to stress, malnutrition, or environmental toxins can impair ANS development, with lasting consequences for physiological regulation. Altered autonomic tone in neonates has been associated with conditions such as sudden infant death syndrome (SIDS) and apnoea of prematurity.

Animal models of autonomic dysfunction provide translational insight into human disorders. Diseases such as familial dysautonomia and Parkinsonian autonomic failure share conserved pathophysiological features across species. Interventional approaches, including vagus nerve stimulation and targeted neuromodulation, show promise in modulating autonomic function, supporting their potential for clinical application.

7.5. Emerging Tools and Future Directions

Technological advancements have profoundly expanded the scope of autonomic neuroscience. Initiatives such as the SPARC programme have yielded rich datasets and high-resolution anatomical atlases, enabling cross-species investigations and refined simulations of autonomic pathways. These resources facilitate both mechanistic understanding and therapeutic innovation.

The integration of genomics, advanced imaging, and computational modelling holds significant promise. Functional MRI in animal models has identified novel circuits implicated in autonomic responses, while machine learning approaches are being employed to detect subtle physiological changes predictive of autonomic dysfunction. Such multidisciplinary strategies are essential for advancing personalised medicine within the domain of autonomic health.

Conclusion

The autonomic nervous system governs a vast array of involuntary processes that are essential for organismal function and survival. Its intricate integration with central and peripheral circuits allows for dynamic adaptation to ever-changing internal and external environments. As research continues to uncover the cellular and systemic complexity of autonomic regulation, translational opportunities expand, offering novel avenues for therapeutic intervention. The future of autonomic neuroscience lies in combining molecular insights with system-level understanding to optimise health outcomes across species.

Chapter 8

Neuroendocrine System – Structure, Function, and Evolution

The neuroendocrine system is a highly integrated network that bridges the nervous and endocrine systems, playing a central role in regulating physiological responses, maintaining homeostasis, and enabling adaptation to environmental change. It comprises interactions between neuropeptides, hormones, and signalling pathways which govern functions such as stress regulation, metabolism, growth, reproduction, and fluid balance. Across animal taxa, this system exhibits both evolutionary conservation and functional specialisation, warranting a detailed examination of its structure, mechanisms, and significance.

8.1. Structure and Components of the Neuroendocrine System

The neuroendocrine system is anchored by the hypothalamic–pituitary axis (HPA), a primary regulatory interface. The hypothalamus synthesises neurohormones, including corticotropin-releasing hormone (CRH) and gonadotropin-releasing hormone (GnRH), which modulate the anterior and posterior pituitary. In response, the pituitary secretes hormones such as adrenocorticotropic hormone (ACTH), luteinising hormone (LH), and oxytocin, which act upon peripheral target organs. This hierarchical organisation permits fine-tuned hormonal control through negative and positive feedback loops.

Neurosecretory neurones are specialised cells that bridge the nervous and endocrine systems. They receive synaptic input, integrate sensory and interneuronal signals, and secrete neurohormones into the circulatory system. In mammals, these cells are located within the hypothalamus and project to the pituitary. In invertebrates, such as Drosophila, converging sensory and interneuronal pathways target neurosecretory cells connected to endocrine tissues. These arrangements support flexible neuroendocrine outputs that respond to internal and external demands.

Neuropeptides such as CRH, vasopressin, and somatostatin are key mediators, influencing processes ranging from stress adaptation to fluid regulation. For instance, Drosophila diuretic hormones (DH44 and DH31) govern fluid balance and nutrient sensing, illustrating functional parallels across phylogenetically distant organisms. These peptides exemplify the precision and diversity inherent in neuroendocrine regulation.

8.2. Mechanisms of Action: Hormonal Regulation and Feedback Loops

Neuroendocrine control relies upon complex feedback mechanisms that maintain physiological stability while ensuring adaptability. In the context of stress, CRH released from the hypothalamus stimulates the pituitary to release ACTH, which in turn promotes glucocorticoid production in the adrenal cortex. These glucocorticoids, such as corticosterone in rodents, enact negative feedback on the HPA axis to modulate and restrain the stress response.

Reproductive function is likewise regulated through pulsatile hormone secretion, a hallmark of neuroendocrine dynamics. In spontaneously ovulating species, ovarian steroids cyclically modulate GnRH release, culminating in an LH surge necessary for ovulation. In induced ovulators, such as rabbits, somatosensory stimulation during mating provokes GnRH release, demonstrating how sensory inputs integrate with hormonal control.

Environmental stressors may also perturb neuroendocrine outputs. Disruptive agents such as neurotoxicants can interfere with receptor function or alter gene expression. Drosophila models have revealed that diuretic hormone pathways adjust fluid regulation in response to ambient carbon dioxide levels, highlighting the system's sensitivity to environmental cues.

8.3. Development and Evolution of the Neuroendocrine System

The neuroendocrine system's developmental trajectory reflects its physiological importance. Neuroendocrine neurones emerge from progenitor populations in the embryonic hypothalamus, guided by transcriptional regulators including Nkx2-1, Fezf2, and Olig2. These cells migrate to defined regions such as the median eminence and establish vascular interfaces essential for hormone release.

From an evolutionary perspective, the neuroendocrine system exhibits deep homology across taxa. Fundamental components, including GnRH and Kisspeptin, are conserved in both vertebrates and invertebrates, indicating common ancestral mechanisms. Although structural features differ, functional principles remain remarkably similar.

Hibernating mammals provide insight into adaptive neuroendocrine evolution. During torpor, neuroendocrine signals—including somatostatin and serotonin—modulate metabolism and thermoregulation, illustrating how conserved systems may be tailored to ecological niches.

8.4. Integration with Other Systems: Neuroendocrine-Immune Crosstalk

The neuroendocrine and immune systems form a bidirectional regulatory network. Immune cells not only express receptors for neurohormones but also secrete peptides that influence neuroendocrine activity. Glucocorticoids, for example, dampen inflammatory responses, while cytokines such as interleukin-1 can alter hypothalamic function and neurohormone release.

This bidirectional communication has profound implications for disease. Dysregulated neuroendocrine–immune interactions are implicated in autoimmune disorders, neuroinflammatory conditions, and neurodegeneration. A detailed understanding of these interactions is essential for the development of targeted therapies that operate at the interface of immunology and endocrinology.

8.5. Stress Response and Behavioural Adaptations

Stress-related neuroendocrine responses are orchestrated through the HPA axis, which adjusts metabolic, immune, and behavioural processes to optimise survival. Glucocorticoids modulate energy availability, inflammation, and mood. Notably, sex hormones influence this axis: oestrogen amplifies, while testosterone attenuates stress reactivity. These interactions exemplify the layered regulation of neuroendocrine stress mechanisms.

Rodent models offer key insights into stress-induced neuroendocrine plasticity. Chronic stress, or prenatal exposure to endocrine-disrupting agents, may induce long-term reprogramming of the HPA axis. Maternal stress during gestation, for instance, is associated with altered offspring neuroendocrine function and increased vulnerability to psychiatric and metabolic disorders.

8.6. Towards a Deeper Understanding of Neuroendocrine Complexity

Despite notable progress, numerous gaps remain in our understanding of neuroendocrine regulation. Future studies should focus on clarifying hormone–receptor interactions, mapping transcriptional cascades governing neuroendocrine development, and examining the enduring impact of environmental disruptors on system function.

Technological innovations are rapidly advancing the field. Techniques such as optogenetics and high-resolution imaging permit real-time tracking of neurohormonal activity and neural circuit dynamics, facilitating a more precise exploration of neuroendocrine processes. Additionally, comparative studies employing model organisms like *Drosophila* and zebrafish offer opportunities to delineate conserved pathways and extend findings to human physiology.

Conclusion

The neuroendocrine system stands as a fundamental regulatory axis, intricately linking neural signals with hormonal outputs to sustain homeostasis and facilitate adaptive responses. Its structural and functional complexity (hypothalamic–pituitary axis, neurosecretory pathways, and diverse neuropeptides) underscores its centrality in governing stress, reproduction, metabolism, and fluid balance.

Mechanistically, the system operates through precisely regulated hormonal feedback loops, which enable rapid and sustained modulation of physiological states. Its ability to integrate sensory inputs with endocrine responses reflects both evolutionary refinement and biological necessity. From the pulsatile secretion of reproductive hormones to the acute reactivity of the stress axis, neuroendocrine mechanisms illustrate the elegant orchestration of internal regulation.

Developmental and evolutionary insights further highlight the conserved nature of this system across species, affirming shared genetic and molecular frameworks. Adaptations observed in unique physiological contexts (hibernation) demonstrate the flexibility of neuroendocrine regulation in the face of ecological demands.

Importantly, the neuroendocrine system does not function in isolation. Its reciprocal communication with the immune system and susceptibility to environmental influences situate it at the crossroads of health and disease. Technological advances continue to deepen our understanding, revealing new therapeutic possibilities while raising important questions regarding environmental disruption and long-term neuroendocrine health.

Moving forward, a deeper mechanistic understanding, informed by cross-species studies and modern methodologies, is essential. Only through such integrative approaches can we fully elucidate the neuroendocrine system's role in both maintaining physiological integrity and responding to the challenges posed by a dynamic internal and external environment.

Chapter 8

Cognitive Function in Animals – Perception, Memory, and Decision-Making

Cognitive function in animals encompasses the processes essential for survival and adaptation. These include perception, learning, memory, and decision-making. These mental abilities are crucial for survival, influencing behaviours such as foraging, mate selection, predator avoidance, and social interaction. While animal cognition was once considered distinct from human intelligence, research reveals significant overlaps, emphasising a continuum of cognitive abilities across species. This section examines key aspects of animal cognition, with a focus on foundational processes, evolutionary insights, and comparative studies.

9.1. Perception and Learning in Animals

Perception forms the foundation of cognitive function, enabling animals to gather and interpret sensory information from their environments. This sensory processing is vital for recognising stimuli, detecting predators, or identifying potential mates. Many animals exhibit remarkable sensory capabilities that are tailored to their specific ecological niches. For example, bats use echolocation to navigate in darkness, while bees perceive ultraviolet light to locate nectar-rich flowers.

Learning complements perception by allowing animals to modify behaviours based on experience. Classical and operant conditioning are fundamental learning mechanisms that enable animals to associate stimuli with outcomes. For example, dogs learn to perform tasks in response to verbal commands, while crows demonstrate insight learning by solving novel problems without prior exposure to them. Advanced learning abilities, such as the use of meaningful communication by parrots, showcase the cognitive diversity among animals.

Evolutionary pressures have shaped learning strategies across species. Carnivores, such as wolves, excel in spatial learning to locate prey, while herbivores focus on recognising safe foraging zones. This variation underscores the interplay between ecological demands and cognitive adaptations, highlighting the evolutionary basis of learning.

9.2. Memory: Retaining and Utilising Information

Memory is a cornerstone of animal cognition, enabling individuals to store, retrieve, and utilise information from past experiences, influencing their ability to adapt to future challenges. Memory in animals can be categorised into two types: episodic-like memory, which involves recalling specific events, and procedural memory, which pertains to skills and routines. These capabilities are not merely passive processes but actively shape future behaviours.

Species with complex social or ecological needs often exhibit exceptional memory capabilities. Elephants, for instance, retain information about water sources across vast territories, demonstrating their ability to maintain long-term spatial memory. Similarly, scrub jays exhibit episodic-like memory, recalling both the location and timing of hidden food caches, which involves sophisticated cognitive processing of spatial and temporal elements.

The neural basis of memory in animals shares similarities with that of humans, with the hippocampus playing a central role in encoding both spatial and episodic memories. Rodent studies have been instrumental in elucidating these mechanisms, providing insights into the molecular and neural processes underlying memory. Understanding animal memory not only sheds light on their adaptive behaviours but also informs research on human memory disorders.

9.3. Decision-Making and Problem-Solving

Decision-making in animals involves evaluating multiple options and selecting actions that maximise fitness. Prior experiences, environmental factors, and social interactions influence this process. Animals make decisions ranging from foraging strategies to navigating complex social dynamics, showcasing the versatility of their cognitive abilities.

Particular species demonstrate remarkable problem-solving skills. For example, chimpanzees use tools to extract food, planning their actions to achieve specific outcomes. Octopuses exhibit similar ingenuity by manipulating objects to access hidden rewards. Even species with smaller brains, such as bees, adapt their foraging behaviour based on resource availability, reflecting advanced decision-making.

Deceptive behaviours highlight another dimension of decision-making. Primates, for instance, have been observed feigning ignorance to avoid sharing food, demonstrating an ability to manipulate social contexts. These behaviours underscore the role of cognition in navigating ecological and social challenges, reinforcing its adaptive value.

9.4. Evolutionary Perspectives on Cognitive Abilities

The evolution of cognitive abilities reflects adaptations to ecological pressures and social demands. Traits such as learning, memory, and tool use have independently emerged across diverse taxa, driven by specific survival and reproductive needs. For instance, tool use in primates and birds has evolved as an innovative solution for accessing food, showcasing convergent evolution in cognitive traits.

Social complexity is a key driver of advanced cognition. The 'social brain hypothesis' suggests that living in groups requires sophisticated mental capabilities, including communication, relationship memory, and predictive behaviour. Dolphins, elephants, and primates exemplify this hypothesis, exhibiting exceptional social intelligence linked to their large brain sizes and intricate social structures.

Cognitive traits are not confined to vertebrates; insects such as ants and bees demonstrate collective intelligence, coordinating complex behaviours through simple rules. These findings challenge traditional notions of cognition, highlighting diverse evolutionary pathways that lead to intelligent behaviour.

9.5. Research Methods and Challenges in Animal Cognition

Advances in methodology have revolutionised the study of animal cognition. Behavioural experiments, such as object discrimination and maze navigation, remain fundamental tools for assessing learning and memory. However, ecological validity, ensuring that tests reflect natural behaviours, is critical for accurate interpretations.

Technological innovations, such as neuroimaging (e.g., fMRI and PET), and computational modelling provide deeper insights into cognitive processes. These methods allow researchers to map brain activity and simulate neural mechanisms underlying cognition. For example, functional MRI studies have identified brain regions involved in decision-making, while computational frameworks, such as the LIDA model, help integrate behavioural and neural data across species.

Despite these advancements, challenges persist. Cognitive tests must account for species-specific differences in motivation and behaviour. Avoiding anthropomorphism, attributing human traits to animals, is essential for maintaining scientific rigour. Addressing these challenges is vital for refining methodologies and advancing the field.

9.6. Comparative Cognition and Human-Animal Continuities

Research on comparative cognition reveals striking parallels between human and animal intelligence. Cognitive processes such as numerical reasoning, concept formation, and self-awareness are not uniquely human but are shared across various species. For instance, pigeons demonstrate numerical cognition, while great apes exhibit a theory of mind, understanding the perspectives and intentions of others.

These findings challenge traditional views of human cognitive superiority, highlighting the evolutionary continuum of intelligence. They also prompt ethical considerations regarding the treatment of animals, given their demonstrated mental capacities. Comparative studies enrich our understanding of cognition, bridging the gap between humans and other species.

Conclusion

The study of cognitive function in animals reveals a rich and diverse landscape of mental abilities that are both adaptive and evolutionarily significant. Processes such as perception, memory, learning, and decision-making are not exclusive to humans but are widely distributed across the animal kingdom, often shaped by ecological pressures and social demands. From the tool use of primates and birds to the navigational memory of elephants and bees, animals demonstrate complex behaviours underpinned by sophisticated cognitive mechanisms.

Advancements in methodology and technology continue to deepen our understanding of these processes, while also highlighting the importance of species-specific approaches and the dangers of anthropomorphism. Comparative studies increasingly reveal cognitive continuities between humans and other animals, challenging anthropocentric assumptions and inviting a more nuanced perspective on intelligence.

As research progresses, the field of animal cognition not only enhances our knowledge of non-human minds but also offers valuable insights into the evolutionary roots of our mental capacities. It compels us to reconsider the boundaries of intelligence, agency, and awareness, ultimately enriching our appreciation of the cognitive lives of the many creatures with whom we share the natural world.

Chapter 10

Pain Perception and Nociception – Mechanisms, Evolutions, and Ethical Considerations

Pain reception and nociception are fundamental survival processes in animals. These mechanisms enable the detection of harmful stimuli, triggering protective responses that are essential for avoiding injury and promoting healing. While nociception refers to the neural processes that detect noxious stimuli, pain is the conscious experience that often arises from these signals. This chapter explores the mechanisms underlying nociception and pain perception, their evolutionary significance, and the methodologies and challenges associated with studying these processes in animals.

10.1. The Mechanisms of Nociception in Animal Neuroscience

Nociception, the neural process of encoding and transmitting noxious stimuli, constitutes the bedrock of pain biology. In animal systems, this mechanism begins with the activation of nociceptors (specialised peripheral sensory neurons) by thermal, chemical, or mechanical stimuli that surpass a biological threshold of potential tissue harm. These primary afferent fibres, notably the Aδ and C fibres in vertebrates, transduce the stimulus into electrical signals which propagate centrally toward the spinal cord or analogous ganglia in invertebrates.

In vertebrates, these inputs typically converge at the dorsal horn of the spinal cord, where second-order neurons integrate, gate, and relay the signals to higher brain regions. Notably, this transmission is not purely passive; neurochemical modulators such as glutamate, substance P, and various neuropeptides amplify or attenuate the signal strength. The complexity increases further in invertebrates, where similar functional responses occur via distinct morphological arrangements, indicating evolutionary convergence rather than homology.

While nociceptive responses can operate without conscious processing, they are nevertheless highly regulated and plastic. Injury-induced sensitisation (a phenomenon common across taxa) illustrates how nociceptive circuits adapt over time, promoting hyperalgesia or allodynia. These alterations, deeply embedded in the neuroethological repertoire of many species, underscore nociception as a conserved survival mechanism rather than merely a prelude to pain.

10.2. Pain Perception: Beyond Sensory Detection

Pain perception involves more than the mere detection of noxious stimuli; it integrates sensory, emotional, and cognitive dimensions. The discriminative-sensory dimension allows animals to localise and assess the intensity of pain, while the affective-cognitive dimension involves emotional and behavioural responses to pain. Neural waves, such as N1, represent the early stages of cortical processing, while later stages, associated with perceptual awareness, are reflected in N2 and P2 waves.

Pain perception, unlike nociception, involves the subjective appraisal of noxious stimuli, typically requiring higher-order processing within the central nervous system. In humans and other mammals, this encompasses cortical structures such as the anterior cingulate cortex, insular cortex, and prefrontal areas—regions known for their roles in affective, evaluative, and motivational dimensions of experience.

Pain can be classified into nociceptive pain, arising from the activation of nociceptors, and neuropathic pain, resulting from damage to sensory fibres or CNS structures. These distinctions are critical for understanding chronic pain conditions and developing effective therapeutic strategies. Research has shown that animals, including fish and mammals, exhibit behaviours indicative of both nociceptive and neuropathic pain, suggesting a shared biological basis for these experiences across species.

Behavioural responses to pain, such as withdrawal reflexes, guarding behaviours, and vocalisations, serve as proxies for assessing pain in animals. The administration of analgesics that mitigate these behaviours further supports the presence of pain perception mechanisms.

In animals, evidence for pain reception (particularly in non-mammalian species) is more elusive. Nonetheless, neurophysiological studies indicate that fish, birds, and cephalopods exhibit central processing responses that exceed simple reflex arcs. These responses include modulated behaviour after injury, self-protective actions, and long-term avoidance learning, all suggestive of an affective state beyond nociception.

One growing area of focus is the dissociation between sensory-discriminative and affective-motivational components of pain. This has prompted the development of experimental paradigms such as conditioned place avoidance and facial grimace scales, both of which infer the presence of pain-like states in animals. While these tools lack the linguistic reporting that defines human experience, they provide robust behavioural proxies for assessing pain in species incapable of verbal expression.

10.3. Neural Pathways of Nociception and Pain Perception

The neuroanatomy of nociception and pain perception reflects evolutionary layering. In vertebrates, the nociceptive signal ascends via the spinothalamic and spinoparabrachial tracts to various brain centres. The thalamus serves as a primary relay, distributing sensory information to somatosensory cortices and limbic structures. This bifurcation underlies the dual nature of pain: its physical localisation and emotional colouring.

In contrast, invertebrates such as crustaceans and insects possess discrete circuits capable of nociceptive plasticity. Although lacking cortical structures, many show modulation at central ganglia, with evidence of sensitisation and avoidance learning. This implies a degree of central integration that, while anatomically distinct from vertebrate brains, performs analogous functions.

Importantly, descending modulation is evident across many taxa. Endogenous opioid systems, long thought to be a mammalian specialisation, have been identified in amphibians, reptiles, and some invertebrates. These descending controls serve to calibrate pain responses to ecological and behavioural contexts, reinforcing the notion that pain systems are evolutionarily adaptive and not merely reactive.

10.4. Evolutionary Perspectives on Nociception and Pain Perception

Nociceptive mechanisms are remarkably conserved across the animal kingdom. From the nematode *C. elegans* to cephalopods and vertebrates, basic features such as TRP channels, sodium ion channels, and neuropeptide signalling are preserved. Studies in insects, molluscs, and fish have revealed behavioural and molecular responses to injury that parallel those in mammals, suggesting a deep phylogenetic origin for nociception.

The presence of such mechanisms in disparate lineages supports the argument that nociception emerged early as a protective adaptation, not as a by-product of complex cognition. The ubiquity of these responses across taxa lends credibility to comparative models in pain research, though care must be taken to distinguish shared function from shared experience.

The evolutionary role of nociception and pain is unmistakably protective. By promoting escape, guarding of injured areas, and learned avoidance, these responses enhance survival. The affective dimension of pain further amplifies its adaptive potency, ensuring prioritisation of injury management even in the face of competing stimuli.

From an evolutionary standpoint, plasticity within pain systems (manifested as sensitisation or desensitization) aligns with environmental demands. In social animals, expressions of pain may also serve communicative roles, eliciting care or caution from conspecifics. These multifaceted functions challenge the traditional dichotomy between reflexive nociception and conscious pain, indicating a spectrum of affective salience across species.

Perhaps the most contested issue in evolutionary pain research is whether invertebrates can truly experience pain. While nociceptive behaviours are well documented in these animals, subjective experience remains speculative. Yet, recent studies in crustaceans, octopuses, and insects suggest a capacity for integrated, flexible behavioural responses following injury, including long-term changes in motivation and decision-making.

Critics argue that such behaviours may still be accounted for by complex reflexes rather than conscious states. However, as evidence mounts for centralised processing, learning, and analgesic sensitivity in invertebrates, the categorical exclusion of these species from ethical consideration becomes increasingly untenable.

10.5. Animal Models for Nociception Studies

Animal models are invaluable for studying nociception and pain, providing insights into the neurobiology of these processes and enabling the development of pain management strategies. Rodents, particularly mice and rats, are widely used due to their physiological similarities to humans and the availability of well-established behavioural tests for nociceptive responses, such as hot plate and tail-flick assays.

Non-mammalian models, including zebrafish and invertebrates like Drosophila and Caenorhabditis elegans, offer complementary advantages. Zebrafish are particularly valuable for high-throughput studies, given their genetic tractability and transparent embryos, which facilitate real-time observation of neural activity. Invertebrate models, with their simpler nervous systems, provide insights into the fundamental genetic and molecular mechanisms of nociception.

Despite their utility, animal models face limitations. For example, the complexity of human pain experiences cannot be fully replicated in animals, and ethical considerations constrain the scope of experimental manipulations. Nevertheless, these models remain indispensable for understanding nociception and the perception of pain.

10.6. Ethical Considerations in Research

Studying pain in animals raises ethical and methodological challenges. Unlike humans, animals cannot verbally communicate their pain, necessitating reliance on indirect measures such as behavioural changes and physiological parameters. This reliance introduces subjectivity and complicates the interpretation of findings.

10.6.1. Animal Welfare

Modern ethical guidelines emphasise the minimisation of pain and distress in animal research. This principle is codified in the Three Rs (replacement, reduction, and refinement) and reinforced through institutional oversight. The use of appropriate anaesthesia, analgesia, and humane endpoints is essential in aligning scientific inquiry with moral obligations.

Particularly in pain research, ethical scrutiny intensifies due to the very nature of the investigations. Studies that deliberately induce pain must demonstrate not only scientific necessity but also careful balancing of harm and benefit. Moreover, as knowledge of animal sentience expands, so too must ethical frameworks evolve to encompass a broader range of taxa.

10.6.2. Assessment Challenges

Inferring pain in animals presents considerable philosophical and methodological challenges. Behavioural proxies, such as withdrawal reflexes or vocalisations, may not reliably reflect affective experience. Furthermore, species differences in expression, physiology, and cognition confound cross-species comparisons.

Technological advances, including neuroimaging and transcriptomics, offer new avenues for assessing pain, but interpretation remains constrained by anthropocentric assumptions. The absence of verbal report does not equate to the lack of experience, and ethical prudence demands erring on the side of caution when evidence is ambiguous.

10.6.3. Animal Welfare

There is growing momentum toward alternatives that reduce animal suffering. These include in vitro systems, computational modelling, and non-invasive techniques. While these approaches cannot yet replicate the full complexity of in vivo systems, they provide valuable adjuncts and can help refine experimental designs.

Additionally, newer behavioural assays such as the facial grimace scale or conditioned place preference offer more nuanced and ethically acceptable ways to assess pain. The refinement of these tools is vital not only for improving scientific accuracy but also for enhancing animal welfare and public trust in research.

Conclusion

Pain and nociception, while intertwined, reflect distinct but complementary processes that lie at the heart of animal neuroscience. Nociception is fundamentally a biological signal (a conserved mechanism that alerts the organism to potential harm). Pain, on the other hand, encapsulates the cognitive and affective interpretation of that signal, demanding complex neural processing and, in many species, perhaps even a form of subjective experience.

Across taxa, from insects to mammals, the evidence reveals a remarkable continuity in molecular components and adaptive behavioural strategies. These systems have evolved not merely to detect danger but to motivate protective action, often modified by learning, context, and internal state. Yet, as our technological reach expands, so too must our conceptual frameworks and ethical boundaries.

The capacity for pain perception in non-human animals, particularly invertebrates, remains a subject of ongoing debate. Nevertheless, emerging behavioural and physiological data call for a more cautious and inclusive ethical stance. In parallel, the refinement of methods, from molecular tools to behavioural assays, strengthens our ability to explore these questions without inflicting unnecessary suffering.

To study pain is to confront both the limits of our knowledge and the boundaries of our moral responsibility. A comprehensive understanding demands more than technical precision—it requires evolutionary insight, interdisciplinary integration, and a commitment to ethical stewardship. In so doing, comparative neuroscience not only advances scientific frontiers but also deepens our respect for the minds and lives of other animals.

Chapter 11

Comparative Neuroscience – Diversity, Evolution, and Insights Across Species

Comparative neuroscience is an interdisciplinary field that examines the diversity of nervous systems across species to understand the principles of brain organization, function, and evolution. By leveraging differences and similarities in neural structures and behaviours, this approach provides invaluable insights into how nervous systems adapt to ecological niches, evolutionary pressures, and developmental constraints. Advances in methodologies, ranging from high-throughput techniques to neuroimaging and molecular tools, have significantly expanded the scope and depth of comparative studies. This chapter explores key aspects of comparative neuroscience, including methodological frameworks, evolutionary insights, and the challenges and potential of this rapidly evolving discipline.

11.1. The Importance of Species Diversity in Neuroscience

The pursuit of understanding how brains give rise to behaviour has long captivated the field of neuroscience. Yet, in recent decades, there has been a stark narrowing of experimental focus to a small set of genetically tractable, laboratory-bound species. While such a shift offers powerful standardisation and technological convenience, it simultaneously obscures the vast biological variability that once informed many foundational discoveries in neurobiology. The call for species diversity in neuroscience is not a nostalgic yearning for methodological pluralism, but a critical imperative to grasp the fundamental and conserved principles governing neural function.

Historically, the field evolved through serendipitous engagements with diverse taxa (from squids and songbirds to crabs and cone snails). These organisms were not selected at random but were instead ideally suited for addressing specific mechanistic questions due to their unique physiological or behavioural features. The comparative approach, as it stands, is not an indulgence but a necessity to avoid epistemological blind spots fostered by over-reliance on a few "model organisms." As we stand on the threshold of transformative genomic technologies, neuroscience is uniquely poised to re-integrate biological diversity into its methodological core.

11.1.1. Broad Insights: Ecological Context, Evolutionary Relevance, and Conservation of Mechanisms

Integrating diverse species into neuroscience research yields insights that transcend superficial anatomical differences. It permits recognition of deep homology (the shared organisational and functional motifs of neural circuits across phyla) while also illuminating species-specific adaptations shaped by ecological demands. For instance, the central complex in insects and the basal ganglia in mammals share striking architectural and functional parallels, suggesting conserved circuitry for behavioural initiation and decision-making despite divergent evolutionary paths.

Species diversity also reveals that behaviour is not simply an output of neural circuitry but a product of interaction between the nervous system and the animal's ecological niche. Neuroethological studies, for example, have shown how behaviours like grooming or prey aversion emerge from complex evolutionary pressures, not always visible in standard laboratory environments. This eco-evolutionary framing underscores that the brain cannot be fully understood outside its natural context.

Furthermore, the comparative approach reinforces translational science. By examining conserved molecular pathways across taxa, such as CREB-mediated plasticity or monoaminergic signalling, researchers gain broader generalisability and robustness in identifying drug targets or biomarkers, especially for psychiatric and neurodegenerative conditions.

11.1.2. Breakthrough Discoveries: The Power of the Unconventional

The history of neuroscience is punctuated by discoveries arising from unexpected species. The ionic basis of action potentials was elucidated using the squid giant axon. Adult neurogenesis was robustly confirmed in songbirds, shifting the dogma of fixed postnatal neural circuits. Studies in turtles have enabled prolonged in vitro analysis of large neural assemblies due to their hypoxia resistance. Insects such as jewel wasps have shown intricate venom-mediated behavioural control in cockroaches, highlighting novel neuromodulatory mechanisms and potential therapeutic avenues.

Moreover, atypical species have unravelled alternative computational principles. Bats, for instance, revealed that grid cell formation in the hippocampus does not require theta oscillations, countering assumptions drawn from rodent models. These revelations fundamentally challenge and refine our conceptual frameworks of neural coding, memory, and spatial navigation.

11.1.3. Limitations of Model Organisms: Bottlenecks, Bias, and Biological Myopia

Despite their utility, model organisms like mice, rats, and zebrafish present profound limitations. Firstly, domestication and inbreeding have drastically narrowed their genetic and behavioural diversity. Traits essential in natural environments, such as complex social hierarchies, predator avoidance, or nuanced emotional states, are often muted or absent, leading to oversimplified interpretations of behaviour and circuitry.

Secondly, model species are often repurposed beyond their biological relevance. For example, efforts to use mice as proxies for human visual perception overlook fundamental sensory and cortical disparities. Such misalignment can skew research priorities and result in translational failures, particularly in fields like neuropsychiatry and neurodegeneration, where mice lack many key pathological features seen in humans.

Finally, reliance on a handful of species fosters a self-reinforcing research culture. The accumulation of tools, funding, and training around established models stifles exploration into potentially fruitful but understudied systems. This bottleneck curtails innovation and risks entrenching artefacts as assumed truths.

11.2. Evolutionary Perspectives and Methodologies in Neuroscience

Modern neuroscience faces a paradox. On one hand, it commands unprecedented technical sophistication and genomic insight; on the other, it risks epistemic constriction through over-reliance on a handful of model organisms and reductionist paradigms. Evolutionary perspectives offer a corrective—a framework grounded in phylogenetic diversity, ecological validity, and an understanding that structure and function in the nervous system are neither universal nor arbitrary, but shaped by adaptive pressures across taxa.

Methodologically, the field has evolved from descriptive neuroanatomy to advanced molecular mapping, yet it has concurrently witnessed a shrinkage in species representation. The genetic revolution, for all its merit, catalysed this narrowing by favouring species amenable to gene editing, breeding, and data integration. While this efficiency has yielded many insights, it has also truncated exploratory scope and obscured lineage-specific solutions to neural computation and behaviour.

To circumvent this bottleneck, evolutionary neuroscience must re-centre methodologies that embrace biological complexity. An integrative framework (drawing upon comparative neuroethology, high-resolution connectomics, and cross-species transcriptomics) promises a broader inference and also a return to the spirit of discovery that animated earlier periods in the field.

11.2.1. Comparative Approaches: Revisiting Evolution's Archive

The comparative method remains one of neuroscience's most potent yet underutilised tools. By examining conserved principles and divergent adaptations across species, comparative neurobiology yields insight into both fundamental mechanisms and organism-specific innovations. The neural basis of motion detection, for example, is instantiated in both fly and mouse using parallel circuit architectures, demonstrating convergence upon effective computational strategies despite vast evolutionary distance.

Likewise, cross-species studies have illuminated deep homology in motor control systems. The structural and functional parallels between the insect central complex and the vertebrate basal ganglia suggest that certain organisational principles—such as the integration of sensory input for behavioural selection—may have ancient origins and be maintained across phyla.

However, the utility of comparative studies extends beyond homology. They reveal how environmental context shapes neural design. Bats, for instance, demonstrate hippocampal spatial coding without theta oscillations, challenging rodent-centric models and underscoring how ecological specialisations yield divergent neural strategies. The field must move beyond tokenistic use of “exotic” species and instead invest strategically in species chosen for their unique evolutionary trajectories, cognitive capacities, or physiological features.

11.2.2. Neuroimaging and Connectomics: Charting Evolution’s Structural Legacy

High-resolution imaging techniques have revolutionised our understanding of the brain’s internal landscape, yet the majority of connectomic data remains constrained to a small cadre of laboratory animals. To capture the breadth of evolutionary solutions to neural architecture, these methodologies must be extended across a phylogenetic spectrum.

Connectomics, when applied comparatively, can reveal which aspects of circuit organisation are conserved and which are labile. Such comparative connectomics can, for example, distinguish universal motifs of sensory processing from species-specific elaborations driven by ecological niches. In the avian brain, despite the absence of a neocortex, structures homologous in function to mammalian cortices mediate complex cognition and learning. This challenges cortico-centric models and encourages reassessment of what constitutes the neural basis of intelligence.

Moreover, recent advances in neuroimaging of freely behaving animals have made it possible to interrogate naturalistic behaviour in ecologically relevant contexts. This shift (from anaesthetised and restrained conditions to dynamic, species-appropriate environments) opens a path to link connectome data with real-time behavioural complexity, thereby restoring ethological validity to the analysis of circuit function.

11.2.3. Genomics and Cell Types: Illuminating Conservation and Innovation

The genomic era provides an extraordinary lens through which to investigate neural diversity. Still, it also presents an ironic hazard—the temptation to assume that molecular homology equates to functional equivalence. While specific transcriptional programmes and signalling pathways, such as CREB-dependent plasticity, are indeed conserved across animal taxa, their phenotypic expression is shaped by a host of regulatory and epigenetic contexts unique to each species.

Transcriptomic comparisons across species reveal that core molecular repertoires involved in learning, memory, and social behaviour are evolutionarily ancient. Yet, the regulatory logic and cellular embedding of these programmes vary substantially. A gene implicated in aggression in mice may contribute to territorial behaviour in sticklebacks or caste differentiation in bees, depending on genomic background and ecological history.

Cell-type classification across species is another frontier of inquiry. Recent work has begun to reveal that cell types previously assumed to be homologous across vertebrates may differ in subtle yet significant ways in their connectivity, electrophysiological profiles, or developmental trajectories. To establish a robust, evolutionarily anchored taxonomy of neural cell types, single-cell transcriptomic studies must be extended to non-traditional organisms, particularly those occupying key phylogenetic positions.

The increasing accessibility of genomic sequencing and gene-editing technologies in non-model organisms promises to dismantle the historical dichotomy between molecular precision and ecological relevance. With CRISPR-Cas and long-read sequencing now deployable in field conditions, the potential for integrated, evolution-aware functional genomics is at hand.

11.3. Challenges in Comparative Neuroscience: Addressing Technological, Conceptual, and Ethical Constraints

Comparative neuroscience, once a fertile ground for discovery, now stands at a crossroads. While the field historically thrived on diversity (leveraging unique species to uncover neural principles), modern neuroscience has converged on a narrow range of model organisms. This shift, propelled mainly by technological convenience and translational pressures, has introduced severe limitations. A comprehensive understanding of the nervous system requires navigating a complex interplay of biological diversity, technological tools, and ethical obligations. Here, we discuss three intertwined challenges in comparative neuroscience: technological bottlenecks, integrative frameworks, and the ethical imperatives of conservation.

11.3.1. Technological Bottlenecks

Technological developments in molecular genetics, imaging, and neuro-engineering have revolutionised neuroscience, but not without cost. Most tools (CRISPR gene editing to in vivo calcium imaging) have been optimised for a few standard models, primarily Mus musculus, Drosophila melanogaster, and Danio rerio. While these organisms offer unparalleled genetic tractability, the cost is a narrowing of biological scope. As Mathuru et al. aptly noted, this overreliance has become a form of intellectual tunnel vision, occluding insights that might be gained from "non-canonical" systems.

The challenge is not merely about access to genomic sequences or transgenic lines, but about translating tools across taxa. Most comparative species lack annotated genomes, optimised viral vectors, or standardised behavioural assays. For instance, studies of higher-order cognition in bats or complex motor control in birds face significant experimental hurdles despite offering clear evolutionary advantages over rodents in specific behavioural domains. Moreover, even when technologies are adapted, the increased phenotypic and genetic variability inherent in wild or outbred species introduces complexity that is difficult to accommodate within current experimental paradigms.

What is required is a technological infrastructure that is both modular and adaptable (one that recognises the trade-off between generality and specialisation). Emerging technologies such as portable optogenetic platforms, miniaturised neural interfaces, and single-cell sequencing for non-model organisms provide glimpses of this future, but their routine use remains aspirational.

11.3.2. Integrative Frameworks

The second major challenge is conceptual: the lack of integrative frameworks that adequately span phylogenetic, ecological, and mechanistic dimensions. Current approaches are predominantly reductionist, excising behaviour from its naturalistic roots and neural circuits from their evolutionary contexts. This has led to models that are mechanistically elegant but ecologically sterile. As Keifer and Summers emphasised, understanding behavioural nuance demands that we return to etho-experimental paradigms and systems-level interpretations.

For example, the convergence in structural design between insect central complexes and vertebrate basal ganglia highlights deeply conserved motifs in motor control. Yet, these findings only emerge when evolutionary homology and functional analogy are both considered. Similarly, the manipulation of behaviour by parasitic wasps or the altered predator response in Toxoplasma-infected rodents exemplify how naturalistic models can reveal neurobiological mechanisms of action selection and behavioural modulation beyond laboratory constraints.

What is needed is a pluralistic and comparative epistemology, one that accommodates the functional specialisation of diverse nervous systems while seeking generalisable principles. This demands collaboration across fields (neuroethology, computational neuroscience, evolutionary biology) and a reappraisal of what constitutes explanatory power in neuroscience.

11.3.3. Conservation and Ethics

Finally, comparative neuroscience bears an ethical responsibility that extends beyond academic pursuits. The continued existence of many species with unique neural adaptations is imperilled by habitat destruction and climate change. The very systems that could offer transformative insights into cognition, sensory processing, and resilience may disappear before they are studied.

There is also an epistemic argument for conservation. Each species embodies a set of evolutionary solutions—genetic, anatomical, and behavioural—to environmental challenges. To lose these systems is to lose possible futures of scientific understanding. Conservation, then, is not merely a moral stance but an epistemological necessity. Brenowitz and Zakon cautioned against the "bottleneck" effect, where the field's entrenchment in a few species risks both stagnation and misrepresentation of biological diversity.

Furthermore, ethical considerations must guide experimental design, especially when venturing beyond traditional laboratory models. Wild-caught animals, for instance, present welfare and ecological concerns that demand rigorous oversight. But such scrutiny should not be a deterrent; rather, it ought to motivate more sensitive, minimally invasive methodologies and collaborative stewardship with conservation biologists.

Conclusion

A robust neuroscience cannot be built upon convenience alone. It must be anchored in intellectual pluralism, embracing a strategic inclusion of diverse animal systems. The value of diversity lies not merely in its breadth, but in its capacity to illuminate both the universal principles of brain organisation and the exceptional adaptations that evolution has crafted. Comparative neuroscience, by its very nature, offers a powerful lens through which we can examine the architecture, plasticity, and function of neural systems across the animal kingdom.

The narrowing of experimental focus to a handful of genetically tractable models has, undeniably, yielded detailed mechanistic insights. Yet this focus comes at a cost: the erosion of evolutionary context and the masking of alternative neural solutions. Emerging technologies (ranging from high-resolution imaging to field-adaptable molecular tools) now render many previously inaccessible species experimentally tractable. This presents a critical opportunity to dismantle the methodological and conceptual bottlenecks that have constrained the field.

Evolutionary perspectives are foundational to neuroscience. They compel us to move beyond the reductive "how" and ask the deeper "why." Why are neural circuits organised in particular ways? Why do certain behaviours persist, and others diverge across taxa? Answering such questions demands a pluralistic methodology integrating comparative anatomy, ethologically grounded behavioural assays, functional genomics, and systems neuroscience within ecologically meaningful frameworks.

To truly understand how brains work, we must engage with the full diversity of life that natural selection has shaped. This means moving beyond the familiar few and turning to the vast spectrum of species whose unique adaptations may hold the key to broader neural principles. In doing so, we not only expand the epistemic reach of neuroscience but also reaffirm our responsibility to conserve the ecological and biological diversity that underpins it.

Comparative neuroscience is thus not a peripheral subfield (it is a vital foundation for a holistic neuroscience of the future). The path forward lies not in deeper excavation of the well-trodden, but in the courageous exploration of the unknown, guided by curiosity, humility, and a commitment to scientific breadth. Only then can neuroscience reclaim its broader vision: one that honours both the unity and the variety of minds across the natural world.

Chapter 12

Social Neuroscience in Animals – The Neural Circuits of Sociality and Communication

Social neuroscience in animals investigates the neural mechanisms that underpin social behaviours, encompassing cooperation, competition, empathy, social learning, and group dynamics. Recent advances have shown that social interactions are mediated by complex, distributed brain networks that encode information about conspecifics, social context, and group structure across a broad range of species, including rodents, primates, birds, fish, and even insects. Key findings include the discovery of inter-brain synchrony during social engagement in rodents, specialised neurones in the primate amygdala and prefrontal cortex that simulate or track the decisions and identities of social partners, and the roles of neuromodulators such as oxytocin and dopamine in modulating social motivation and neural plasticity.

Methodological innovations, such as simultaneous multi-brain recordings and ethologically relevant behavioural paradigms, have enabled more naturalistic investigations of social behaviour. Nevertheless, challenges remain in translating findings across species, capturing the full complexity of social interactions, and understanding the impact of social isolation or early-life experiences on neural circuits. This review synthesises the latest research, highlighting both foundational discoveries and emerging themes within the field of animal social neuroscience.

12.1. The Neural Mechanism of Social Behaviour

Social behaviour in animals arises from intricate neural processes that integrate sensory inputs, internal states, and contextual cues to guide interactions with conspecifics. Recent interdisciplinary advances in neuroscience, particularly combining behavioural paradigms with electrophysiology, optogenetics, and neuroimaging, have identified a distributed network of brain regions underlying social cognition, motivation, and action across species (from rodents to non-human primates).

12.1.1. Prefrontal Cortex and Interbrain Synchrony

The prefrontal cortex (PFC) plays a pivotal role in regulating social cognition and decision-making. In rodents, studies using dual recordings in socially interacting pairs have demonstrated interbrain synchrony, particularly in the medial prefrontal cortex (mPFC), that predicts social dominance and affiliative behaviours. Kingsbury et al. (2019, Cell) showed that pairs of mice engaged in social encounters exhibited temporally coordinated activity in the mPFC, and that this synchronisation varied depending on the dominance relationship. Disruption of this synchrony through chemogenetic inhibition impaired the stability of social hierarchies, suggesting a causal role in modulating dominance dynamics.

12.1.2. Hypothalamus and Amygdala

The Core of Social Motivation and Risk Processing, the hypothalamus, particularly the medial preoptic area (mPOA) and ventromedial hypothalamus (VMHvl), has been implicated in sexually dimorphic and aggression-related behaviours. For example, optogenetic stimulation of the VMHvl in mice triggers aggressive attacks, while inhibition reduces aggression. These findings highlight the hypothalamus's role in encoding social motivation and threat perception.

Meanwhile, the amygdala integrates emotional salience and social valence. In both rodents and primates, specific populations of amygdala neurones respond selectively to facial expressions, dominance cues, and social context. Amygdala neurons that respond to eye contact and facial identity, suggesting their role in social recognition and emotional evaluation.

12.1.3. Hippocampus and Striatum

The hippocampus, classically associated with spatial memory, also supports social memory (the ability to remember individuals and their social status). The ventral hippocampal CA1 neurons in mice have been known to encode social memories, and optogenetic silencing of these cells impairs the ability to distinguish familiar from novel conspecifics.

The striatum, particularly the nucleus accumbent, integrates social reward information and supports motivated social behaviours. Dopaminergic projections from the ventral tegmental area to the nucleus accumbent modulate social interaction preferences in rodents, providing evidence for a reward circuitry that is tuned to social stimuli.

12.1.4. Primate Studies: Simulation and Social Encoding

In non-human primates, the orbitofrontal cortex (OFC) and anterior cingulate cortex (ACC) have been shown to encode social value and partner choices during strategic decision-making tasks. Single neurons in the macaque ACC gyrus have been known to respond whether a reward was delivered to the self, another monkey, or both. This provides direct evidence of neural coding of other-oriented decisions, a foundation for empathic behaviour and social cooperation.

Moreover, the amygdala and PFC together form a circuit that simulates the intentions and decisions of social partners. The amygdala neurons in macaques have also been known to predict the future choices of social partners, suggesting a mechanism for mentalising or theory of mind-like processing.

12.2. Interbrain Coupling and the Dynamics of Joint Decision-Making

Recent methodological advancements have enabled simultaneous neural recordings from multiple socially interacting individuals, offering a granular perspective on how brains coordinate during real-time exchanges. In rodents, the dorsomedial prefrontal cortex (dmPFC) has emerged as a critical hub in facilitating interbrain synchrony, a phenomenon wherein activity across individuals becomes temporally aligned during interaction. This synchrony arises not from passive sensory co-experience but from active, reciprocal social engagement.

Interbrain neural synchrony in freely interacting mice predicted subsequent social behaviour and dominance hierarchies. Through dual microendoscopic calcium imaging of dmPFC populations, they identified distinct neuronal ensembles encoding both self-generated and partner-generated behaviours. This division of labour at the neuronal level suggests a distributed coding scheme for integrating one's actions with the inferred states and decisions of others. Crucially, when physical barriers precluded interaction, synchrony collapsed, confirming that neural coupling is contingent upon dynamic engagement, not merely shared environmental stimuli.

In competitive contexts, such as tube dominance tests, interbrain coupling remained robust and predictive of dominance outcomes. Dominant mice displayed a higher frequency of proactive social behaviours, while subordinates showed increased reactive retreating. Despite these behavioural asymmetries, their dmPFC activity remained synchronised, indicating that synchrony may not solely reflect cooperative alignment, but rather a broader neurobiological mechanism for tracking and reacting to others in real-time competitive dynamics.

12.3. Social Prediction and Theory-of-Mind-Like Computation in Non-Human Primates

While rodents provide tractable models for dissecting circuit-level mechanisms, primate studies offer invaluable insights into higher-order cognitive functions that parallel human social reasoning. There is a novel paradigm in which macaques engaged in a repeated iterated prisoner's dilemma (iPD) game, allowing for the exploration of cooperation and defection under varying social contingencies. Using single-unit recordings from the dorsal anterior cingulate cortex (dACC), the researchers identified neurons that predicted an opponent's future decision before it was made, a finding that echoes the conceptual foundation of the theory of mind.

Significantly, the activity of these "other-predictive neurons" was not confounded by expected reward or the monkey's own action, indicating a dedicated encoding for the covert decisions of others. Mixed neuronal populations within the dACC could decode the future cooperative intentions of the partner at near-optimal levels, closely mirroring behavioural performance of the animals. Moreover, when dACC activity was transiently disrupted, monkeys showed diminished propensity for mutual cooperation, without impairing individual decision-making in non-social contexts. This result underscores the causal necessity of dACC activity in enabling mutually beneficial social behaviour.

12.4. Identity Encoding and Multi-Agent Social Cognition

Expanding from dyads to group interactions uncovered an even more sophisticated form of social coding in the macaque prefrontal cortex. Their triadic decision-making paradigm required individuals to choose between rewarding one of two peers, revealing the necessity for agent-specific identity representation in real-time strategic decision-making. Through recordings from the dorsomedial prefrontal cortex, they discovered cells that selectively encoded not only the identity of social partners but also the nature of prior interactions, including reciprocity and retaliation.

These "social agent identity cells" supported complex behavioural strategies such as tit-for-tat, whereby animals reciprocated reward offers from specific individuals while retaliating against perceived exploitation. Such cell populations integrated actor-recipient dyads, past outcomes, and future strategy, functioning as a neural substrate for encoding abstract social relationships within groups. Notably, behavioural strategies were not merely spatially driven or based on conditioned cues; animals responded flexibly even when partner locations were swapped, affirming that identity representations were intrinsic and context-dependent.

In contrast to canonical mirror neurons, which generalise action observation, these agent-identity cells may support person-specific encoding, enabling the flexible formation of alliances and the mitigation of exploitation. This coding principle offers a cellular resolution account of how animals track social histories and navigate group dynamics, which may be evolutionarily conserved across complex social species.

12.5. Genetic and Molecular Correlates of Social Hierarchies

While neural circuits underpinning social behaviour have been well characterised at the level of brain activity and regional specialisation, an emerging body of evidence suggests that social status is not merely a function of circuit dynamics but also reflects molecular and genetic adaptations, as gene expression in the brains of group-living mice was directly correlated with stable patterns of social dominance. This work elegantly bridges behavioural ethology, social network theory, and molecular neuroscience, offering a multiscale account of how hierarchy is established and biologically sustained.

In their experimental framework, So and colleagues tracked twelve outbred CD1 male mice housed in a large three-dimensional vivarium for 21 days. Their behavioural observations captured more than 100 hours of spontaneous social interactions, encompassing agonistic behaviours (e.g., fighting and chasing), as well as affiliative and investigatory interactions such as allogrooming and anogenital sniffing. These continuous observations enabled the construction of weighted social network matrices describing dyadic interaction frequencies and directions, from which dominance ranks and network centralities were derived using both classical and novel analytical methods.

One of the more innovative aspects of the study was the application of Glicko rating systems (a dynamic approach borrowed from chess ranking algorithms) to infer each individual's changing social status over time. This method revealed that dominance hierarchies were not only steep and linear but also temporally stable. Importantly, dominance was not a transient behavioural state; rather, it emerged as a robust trait consistent across time and measurable via quantitative network metrics such as Kleinberg's Hub Centrality and Bonacich's Power Centrality.

Upon establishing these hierarchies, the neurogenetic substrates that might support or reflect an animal's position in the social order are investigated. Using in situ hybridisation, they quantified gene expression in brain regions canonically associated with social cognition: the medial and central nuclei of the amygdala, the medial preoptic area of the hypothalamus, and the hippocampus. Four key molecular markers were analysed:

12.5.1. Corticotropin-Releasing Factor (CRF) and the Neuroendocrine Substrates of Dominance

CRF plays a central role in the hypothalamic–pituitary–adrenal (HPA) axis, orchestrating stress responses and modulating socio-emotional behaviour. In dominant mice, CRF mRNA expression was significantly elevated in the medial and central amygdala, as well as in the medial preoptic area of the hypothalamus, when compared to subordinates. This expression profile suggests that dominant individuals may exhibit enhanced neuroendocrine responsivity, potentially facilitating rapid behavioural adaptability during social contests.

These findings align with broader evidence implicating CRF in social vigilance, arousal, and behavioural inhibition, which are essential traits for maintaining social status. Moreover, CRF's role in the preoptic hypothalamus links it to reproductive and parental behaviours, suggesting that dominance may be evolutionarily intertwined with reproductive opportunity and investment.

12.5.2. Brain-Derived Neurotrophic Factor (BDNF)

Brain-derived neurotrophic factor (BDNF), a critical modulator of synaptic plasticity and neuronal survival, was elevated in the hippocampus of dominant individuals, particularly within the dentate gyrus and CA1 regions. As the hippocampus is well established in encoding contextual memory and individual recognition, increased BDNF expression may underpin the enhanced social memory and learning capacities required by dominant individuals to monitor rivals, track social exchanges, and adapt strategies accordingly.

Strikingly, prior studies have shown that BDNF is not only upregulated by social interaction but is also sensitive to social enrichment or deprivation, providing a potential molecular mediator by which early social experiences can shape adult hierarchy status. Dominant individuals may thus exhibit proactive engagement in social learning, facilitated by heightened hippocampal plasticity.

12.5.3. Glucocorticoid Receptor (GR)

The expression of glucocorticoid receptors (GR) in the hippocampus was also positively correlated with dominance rank. GRs modulate negative feedback in the HPA axis, thereby regulating stress responsiveness and behavioural inhibition. Elevated GR levels in dominant mice suggest a more finely tuned regulatory control of stress reactivity, which could confer an advantage in maintaining status without incurring chronic physiological costs.

Importantly, this result contrasts with findings in some primate species, where subordinates display higher cortisol levels, often interpreted as a reflection of psychosocial stress. The discrepancy may be due to differences in species-specific social organisation, environmental predictability, or contextual controllability. In the more dynamic and conflict-prone mouse hierarchies, dominance may actually select for or result in neuroendocrine flexibility, rather than sustained suppression or activation of the stress axis.

12.5.4. Integrative Implication

Taken together, these findings reinforce the notion that social dominance is not simply behavioural but involves coordinated molecular adaptations that span multiple brain regions and signalling pathways. The stable expression patterns observed across dominant individuals indicate that genomic regulation is integral to the expression and maintenance of social status.

Moreover, this study illustrates the utility of social network analysis in neurogenomic research. Rather than relying on binary classifications of 'winner' or 'loser,' the integration of continuous, multivariate network metrics with gene expression profiling allows for a more nuanced understanding of individual differences. It opens the door to future investigations into how social experience modulates gene expression, or conversely, how pre-existing molecular traits might predispose individuals to ascend or descend within a social hierarchy.

It is also noteworthy that social network topologies differed significantly between behaviours: while grooming was relatively unstructured and independent of dominance, sniffing networks were highly predictive of agonistic directionality, and thus more informative about social structure. This divergence in behavioural modules underscores the importance of multi-dimensional phenotyping when assessing social traits and their biological correlates.

Conclusion

Social behaviour in animals emerges from the seamless integration of neural circuits, behavioural dynamics, and molecular underpinnings, collectively shaping how individuals perceive, evaluate, and respond to one another within diverse ecological and social contexts. This chapter has elucidated the complexity of social cognition by tracing its neural foundations across species, scales, and systems.

At the circuit level, we observe a convergence of distributed brain regions (most notably the prefrontal cortex, amygdala, hippocampus, and striatum) that dynamically encode social value, identity, memory, and reward. Interbrain synchrony, particularly in rodents, highlights the temporal coordination of neural activity during real-time social interactions, reflecting not just mutual awareness, but an active co-construction of behavioural outcomes such as dominance and cooperation. In non-human primates, specialised neural populations simulate the intentions of others, supporting theory-of-mind-like computation and enabling adaptive, strategic behaviour within fluid social hierarchies.

Expanding beyond dyadic interactions, the discovery of agent-specific encoding and triadic social decision-making circuits reveals that animals are capable of forming nuanced social representations. These neural architectures allow for reciprocity, alliance formation, and the encoding of social histories, capabilities once thought to be uniquely human.

At the molecular level, social status and behavioural tendencies are mirrored in the expression of key neuroendocrine and neuroplasticity-related genes, such as CRF, BDNF, and glucocorticoid receptors. These molecular profiles are not static but modulated by social experience, suggesting a powerful bidirectional relationship between social environment and gene regulation. Such findings imply that sociality is not only embedded in neural computation but also biologically embodied at the genomic level.

Importantly, this chapter underscores the value of methodological innovation (multi-brain recordings, optogenetics, dynamic social paradigms, and social network analysis), which has enabled more ecologically valid models of animal sociality. These approaches are reshaping how we understand the biological roots of cooperation, competition, empathy, and communication.

In sum, animal social neuroscience offers a compelling framework to study the evolutionarily conserved substrates of complex social behaviours. It bridges levels of analysis from molecules to groups, providing critical insights not only into the nature of social cognition but also into its disruption in neuropsychiatric and developmental disorders. As research continues to push the boundaries of what can be observed and manipulated, we are poised to uncover even deeper principles that govern social life.

Chapter 13

Neuroscience of Social Stress – Neural and Endocrine Mechanisms of Social Threat, Isolation, and Resilience

Social stress is among the most potent and pervasive challenges faced by social animals. From dominance hierarchies in rodents to peer rejection in primates and loneliness in humans, social threats and disruptions have measurable impacts on brain structure, function, and behaviour. The field of social stress neuroscience explores how social environments, especially those involving threat, exclusion, or instability, will shape neural circuits, endocrine systems, and long-term health outcomes.

This chapter examines the neural, molecular, and hormonal mechanisms of social stress, focusing on evidence from animal models and their translational relevance to human mental health.

13.1. Defining Social Stress: Conceptual Frameworks

Social stress encompasses a wide array of conditions: subordination, isolation, overcrowding, unstable hierarchies, social defeat, rejection, or the threat thereof. Unlike physical stressors, social stress involves perceived or anticipated threats to social bonds, status, or safety, conditions that have profound relevance across species.

Social stress can be acute (e.g., a single defeat or exclusion event) or chronic (e.g., prolonged isolation or repeated subordination). In both cases, animals show robust behavioural, neurochemical, and immunological changes, with the chronic forms posing a greater risk for long-term dysfunction.

13.2. Neural Circuits of Social Threat Processing

13.2.1. Amygdala and Threat Detection

The amygdala plays a pivotal role in detecting socially salient stimuli, particularly those signalling threat or exclusion. In rodents and primates, amygdala neurons respond preferentially to dominance cues, aggressive postures, facial expressions, and signs of social defeat. In socially stressed animals, amygdala hyperactivity is frequently observed and correlates with elevated anxiety-like behaviour.

13.2.2. Prefrontal Cortex and Emotional Regulation

The medial prefrontal cortex (mPFC), particularly the infralimbic and prelimbic regions in rodents and the ventromedial prefrontal cortex in primates, is critically involved in evaluating social threat and regulating stress responses. Chronic social stress is associated with: Reduced mPFC activity and volume, impaired regulation of the hypothalamic–pituitary–adrenal (HPA) axis, and decreased cognitive flexibility.

In primates, dorsal anterior cingulate cortex (dACC) activity also tracks social exclusion and rejection, indicating that these experiences are processed as emotionally aversive and engage pain-related circuits.

13.2.3. Hypothalamus and Aggression Circuits

The ventromedial hypothalamus (VMHvl) is involved in encoding social aggression and threat, especially in contexts of territoriality or dominance contests. Chronic activation of VMHvl-related circuits can reinforce aggression-prone phenotypes, while their inhibition may blunt social vigilance and responsiveness.

13.3. The HPA Axis and Social Stress Hormones

One of the most well-characterised effects of social stress is the activation of the HPA axis, resulting in the release of glucocorticoids (cortisol in humans and primates; corticosterone in rodents). These hormones facilitate short-term adaptation but may be maladaptive when chronically elevated.

13.4. Social Defeat and Isolation Paradigms

13.4.1. Chronic Social Defeat Stress (CSDS)

The chronic social defeat stress model in mice, where a subject is repeatedly exposed to aggression from a larger, dominant conspecific, has become a powerful paradigm to study social stress-induced depression. Mice subjected to CSDS show:

a. Social avoidance
b. Anhedonia
c. Altered dopaminergic signalling
d. Reduced neurogenesis

CSDS induces long-lasting changes in the nucleus accumbent, ventral tegmental area, and mPFC. These changes recapitulate key features of human depression and PTSD, making the model highly translational.

13.4.2. Social Isolation and Loneliness

Social isolation during critical developmental windows, such as post-weaning or adolescence, leads to:

a. Reduced BDNF in the hippocampus
b. Impaired myelination in the prefrontal cortex
c. Social recognition deficits
d. Elevated anxiety and stereotypies

In adult animals, prolonged isolation affects interbrain synchrony, reduces social preference, and compromises mating behaviour. In humans, chronic loneliness has been associated with:

a. Increased mortality risk
b. Accelerated cognitive decline
c. Immune dysregulation

13.5. Neuroplasticity and Resilience

Not all individuals exposed to social stress succumb to maladaptive outcomes. Resilience is an emerging concept defined by the capacity to maintain or recover normal physiological, behavioural, and cognitive function despite adversity. In recent years, studies have begun to uncover the neural correlates of resilience, particularly in animals that resist developing anxiety- or depression-like behaviours after chronic social defeat or isolation.

Resilient animals often exhibit adaptive coupling between the medial prefrontal cortex (mPFC) and the amygdala, allowing for more effective regulation of emotional responses to social threat. Additionally, they demonstrate preserved hippocampal neurogenesis, supporting memory and contextual processing, as well as a balanced dopaminergic tone in mesolimbic pathways, which sustains motivation and reward sensitivity. At the molecular level, transcriptional profiling in resilient phenotypes reveals an upregulation of anti-inflammatory genes, increased expression of neurotrophic factors, and enhanced support for mitochondrial function, all of which contribute to cellular robustness under stress.

Efforts to promote resilience have identified several modifiable factors. Environmental enrichment, voluntary physical exercise, and predictable, socially supportive environments are known to enhance resilience markers and behavioural flexibility. Furthermore, epigenetic modulators, such as histone deacetylase (HDAC) inhibitors, show promise in reversing gene expression changes associated with social defeat, thereby restoring synaptic plasticity and stress buffering capacity. Behavioural training that encourages prosocial interaction, including structured group housing or affiliative learning tasks, has also been shown to reinstate social motivation and restore neural function in previously stressed individuals.

13.6. Translational Relevance: Human Health and Policy

The mechanisms uncovered in animal models of social stress have clear analogues in human psychiatric conditions. Experiences such as social isolation, bullying, workplace harassment, and relational trauma are now recognised as potent drivers of major depressive disorder, generalised anxiety, post-traumatic stress disorder (PTSD), and even suicidality. Rather than viewing these outcomes solely as individual psychological failures, a neuroscience-informed perspective highlights their roots in dysregulation of social neural circuits, including those governing threat detection, emotion regulation, and social cognition. This reconceptualisation has important implications for intervention. It calls for a shift in priorities towards preventative mental health strategies within schools and communities, the development of public health frameworks that foster social cohesion and inclusion, and the implementation of workplace policies that safeguard psychological safety and promote social predictability.

Conclusion: Toward a Socially-Aware Neuroscience

Social stress is not an ancillary concern within neuroscience; rather, it lies at the heart of understanding the intricate interface between brain, behaviour, and environment. The human brain evolved to navigate complex social landscapes, and many forms of dysfunction emerge when these interactions break down or become chronically disrupted. As the field advances, neuroscience must increasingly reckon with the contextual and collective nature of cognition, recognising that brains do not operate in isolation, neither biologically nor socially.

Future research directions should prioritise longitudinal tracking of social experience and neural development, the integration of social network metrics with brain imaging, and cross-disciplinary collaboration with fields such as sociology, anthropology, and behavioural economics. In aligning neuroscience with humanistic and public health goals, we pave the way for a more responsive, inclusive, and socially literate science, one that acknowledges and honours the profound neural architecture underpinning our social nature.

Moreover, embracing a socially aware neuroscience means acknowledging that environmental inequities, social exclusion, and chronic interpersonal stress are not merely psychosocial phenomena, but also exert measurable and enduring effects on brain structure and function. Social determinants of health must therefore be treated as integral to neurobiological research, not peripheral to it. This perspective encourages scientists, clinicians, and policymakers to adopt an explicitly relational framework, one that promotes equity, inclusion, and dignity in all aspects of mental health care, education, and community life.

Ultimately, by integrating the insights of social neuroscience into broader scientific, clinical, and societal frameworks, we move toward a model of brain science that not only seeks to understand neural systems but also to improve the social worlds in which those systems flourish or fail.

Chapter 14

Epilogue – The Social Brain and the Future of Neuroscience

As this volume draws to a close, one truth resonates across every chapter: the brain is inherently social. From the fine-tuned circuits that govern recognition and affiliation to the complex neuroendocrine responses to hierarchy and isolation, sociality is not merely a behavioural trait, it is embedded in the very architecture of the nervous system. The scientific insights presented in this book illustrate, across taxa and methodologies, that neural systems are not solitary processors of information, but dynamic participants in a continuously unfolding social world.

Whether in birds, rodents, fish, or primates, the brain emerges not as a passive receiver of environmental input but as an organ shaped by the challenges of group living: cooperation, competition, empathy, alliance formation, and social learning. From this perspective, even the most individualistic of behaviours, decision-making, memory consolidation, and emotional regulation, are cast in a new light: as products of systems evolved to manage interpersonal complexity, to anticipate the actions of others, and to maintain adaptive relationships across dynamic social landscapes.

14.1. Sociality as a Central Theme in Brain Evolution

Evolutionary pressures have sculpted brains that are remarkably sensitive to social structure and capable of responding flexibly to shifts in social context. In species that rely on group cohesion for survival, such as primates, elephants, cetaceans, and certain birds, social intelligence has co-evolved with brain regions associated with executive function, memory, and affective regulation. This evolutionary trajectory underlies what has been termed the "social brain hypothesis," proposing that increasing neocortical volume in some species correlates more strongly with social group size and complexity than with ecological variables alone.

In this light, understanding the neural basis of social cognition is not merely a subfield of neuroscience, it is a foundational lens through which the brain itself may be understood. The discoveries discussed in this book, from interbrain synchrony in rodents to social reward circuits in primates, reinforce this claim. They show that the core principles of social neuroscience, such as the encoding of conspecific identity, the computation of social value, and the regulation of social threat are deeply conserved across phylogeny, offering a comparative framework for unravelling the neural correlates of human social behaviour and dysfunction.

14.2. From Neural Circuits to Societal Challenges

The translational implications of this work are substantial. As detailed in the latter chapters, disruptions to social circuitry are increasingly implicated in a wide range of neuropsychiatric and neurodevelopmental conditions, including autism spectrum disorder, schizophrenia, depression, anxiety disorders, and frontotemporal dementia. These conditions are often accompanied by profound alterations in social perception, affiliation, and interaction, features that frequently determine prognosis, caregiver burden, and quality of life.

Significantly, the field has moved beyond mere observation of correlation. Through carefully designed animal models, researchers can now demonstrate causal links between molecular, genetic, and circuit-level changes and specific social behaviours. For instance, chronic social defeat models in rodents have illuminated how repeated social stress alters reward processing, increases inflammation, and suppresses neurogenesis, biological changes also observed in human mood disorders. Similarly, disruptions to interbrain synchrony in the prefrontal cortex have been proposed as neural signatures of impaired social reciprocity, a hallmark of autism spectrum conditions.

By investigating these phenomena in animals, scientists can probe mechanisms in ways that are ethically and practically impossible in human populations. This mechanistic clarity enables targeted interventions, such as optogenetic modulation of specific circuits, the testing of neuromodulatory agents like oxytocin and vasopressin, or the reversal of stress-induced plasticity through environmental enrichment and behavioural rehabilitation.

14.3. A Call for Interdisciplinarity

Yet, the brain does not exist in isolation, and neither should its study. One of the most pressing needs in contemporary neuroscience is greater interdisciplinary collaboration. To fully understand the implications of neural sociality, we must engage with insights from sociology, anthropology, psychology, behavioural ecology, and public health. These fields offer critical perspectives on how social environments are constructed, how norms and hierarchies are enforced, and how marginalisation or inclusion affect health and development.

Indeed, the biological is inseparable from the social. Neural systems may encode threat or reward, but it is social experience of inclusion, rejection, power, or empathy, that shapes their tuning. A socially aware neuroscience must therefore acknowledge the diversity of social structures across cultures, the plasticity of social roles over the lifespan, and the profound effects of social determinants on brain health.

To this end, cross-species research remains invaluable. By identifying commonalities and divergences in social brain organisation, we not only enhance our understanding of evolutionary pressures but also identify shared vulnerability and resilience factors in the face of social adversity. For instance, both humans and non-human primates show stress-related cortisol elevations in unstable hierarchies, suggesting convergent physiological responses to perceived social threat.

14.4. Relevance to Global Health and Development

The relevance of social neuroscience extends far beyond the clinic or laboratory. It intersects directly with several United Nations Sustainable Development Goals (SDGs), particularly:

SDG 3: Good Health and Well-being – By illuminating how early-life social environments shape brain development, and how social isolation or threat contributes to mental illness, neuroscience can inform more effective, preventive public health strategies.

SDG 10: Reduced Inequalities – Neuroscience must also address structural barriers to health, recognising that systemic inequality and exclusion are not merely political or economic issues but deeply neurobiological ones.

A socially-attuned neuroscience can contribute to education policy, advocating for emotionally safe classrooms that support social learning and peer attachment. It can guide urban planning, suggesting ways to design communities that reduce loneliness and promote social cohesion. And it can inform workplace design, recommending structures that balance collaboration with autonomy, and predictability with innovation.

14.5. Towards a Socially Literate Science

Looking forward, several priorities should shape the future of this field. First, longitudinal studies are needed to trace how early social experiences interact with genetic predispositions to influence neural development across the lifespan. Second, the integration of social network analysis with brain imaging and computational modelling offers promising new tools for understanding how individuals operate within collective dynamics. Third, researchers must prioritise diversity and inclusivity, both in terms of participant populations and theoretical frameworks, to avoid reproducing narrow conceptions of what constitutes 'normal' social functioning.

Equally important is the need to communicate findings in ways that are accessible, actionable, and ethically grounded. As neuroscientific knowledge becomes increasingly influential in shaping policy and practice, it must be wielded with humility and responsibility. We must avoid biological determinism and instead emphasise the plasticity and contextuality of brain function. In doing so, we reinforce the message that neural circuits are not fixed fates but dynamic reflections of lived experience.

14.6. Final Reflections

Ultimately, this book affirms that to understand the brain is to understand connection. The synapse, the network, the circuit (these are not metaphors for social life; they are its biological foundation). From fish to primates, from solitary vigilance to group synchrony, from maternal bonding to coalition politics, the nervous system reveals its evolutionary heritage through the language of sociality.

The central lesson from comparative social neuroscience is that brains are not solitary organs. They evolve in social niches, adapt to group-specific challenges, and function optimally within a context of reciprocal relationships. The neural substrates of attachment, empathy, aggression, and communication do not operate in a vacuum; they are embedded in environments rich with history, emotion, and meaning. As such, understanding the nervous system in full demands a contextual, culturally sensitive, and socially literate approach.

In this closing reflection, we urge the scientific community to continue building a neuroscience that is not only rigorous but also responsive to the human condition. A neuroscience that values both the neuron and the narrative, both the data point and the dignity of the subject. In doing so, we move toward a science that is as socially intelligent as the brains it seeks to understand.

This vision of neuroscience invites researchers, clinicians, and educators to bridge the gap between bench and society. Laboratory models, as precise and controlled as they are, must be complemented by insights from lived experience. In animals, as in humans, the richness of social life (including affection, conflict, ritual, and memory) cannot be reduced to neural firing rates alone. It must be appreciated through a multilevel lens that incorporates physiology, behaviour, context, and development.

Furthermore, the ethical implications of social neuroscience demand renewed attention. As we uncover more precise ways to decode social cognition, influence trust, or modulate affiliation, we must be vigilant against misuse. The potential to alter or commodify social behaviour, through pharmacology, surveillance, or predictive algorithms, is real. A socially responsible neuroscience must thus include a bioethical framework that upholds individual autonomy, cultural diversity, and relational dignity.

We also call for an educational reorientation. Students of neuroscience should be equipped not only with technical acumen but with an appreciation of the interpersonal dimensions of cognition and emotion. Cross-disciplinary training that integrates anthropology, ethics, sociology, and philosophy can cultivate a generation of scientists who are not only skilled but also wise, capable of navigating both the biological intricacies and the moral responsibilities of their work.

As we look ahead, let us not lose sight of the central revelation this book has traced: that our relationships sculpt our brains, and that neural health is inextricable from social wellbeing. This principle carries implications not only for research but for public policy, health systems, education, and community design. In every arena where brains develop, mature, and age, the quality of social connection will remain a fundamental determinant of thriving.

Let this be the legacy of social neuroscience in animals: not merely a catalogue of circuits or behaviours, but a scientific invitation to understand connection as the essence of life, and the brain as its most eloquent expression.

References

Acar, F., Naderi, S., Guvencer, M., Türe, U., & Arda, M. (2005). Herophilus of Chalcedon: a pioneer in neuroscience. *Neurosurgery*, 56 4, 861-7; discussion 861-7. https://doi.org/10.1227/01.neu.0000207965.83692.9b.

Alvarez-Bolado, G. (2018). Development of neuroendocrine neurons in the mammalian hypothalamus. *Cell and Tissue Research*, 375, 23 - 39. https://doi.org/10.1007/s00441-018-2859-1.

Alves, P., Foulon, C., Karolis, V., Bzdok, D., Margulies, D., Volle, E., & De Schotten, T. (2019). An improved neuroanatomical model of the default-mode network reconciles previous neuroimaging and neuropathological findings. *Communications Biology*, 2. https://doi.org/10.1038/s42003-019-0611-3.

Arakawa, H., & Iguchi, Y. (2018). Ethological and multi-behavioral analysis of learning and memory performance in laboratory rodent models. *Neuroscience Research*, 135, 1-12. https://doi.org/10.1016/j.neures.2018.02.001.

Arbilly, M., & Lotem, A. (2017). Constructive anthropomorphism: a functional evolutionary approach to the study of human-like cognitive mechanisms in animals. *Proceedings of the Royal Society B: Biological Sciences*, 284. https://doi.org/10.1098/rspb.2017.1616.

Arshavsky, Y. (2017). Neurons versus Networks: The Interplay between Individual Neurons and Neural Networks in Cognitive Functions. *The Neuroscientist*, 23, 341 - 355. https://doi.org/10.1177/1073858416670124.

Atta, H. (1999). Edwin Smith Surgical Papyrus: The Oldest Known Surgical Treatise. *The American Surgeon*, 65, 1190 - 1192. https://doi.org/10.1177/000313489906501222.

Attaai, A., Noreldin, A., Abdel-Maksoud, F., & Hussein, M. (2020). An updated investigation on the dromedary camel cerebellum (Camelus dromedarius) with special insight into the distribution of calcium-binding proteins. *Scientific Reports*, 10. https://doi.org/10.1038/s41598-020-78192-7.

Austen, I. (2021). Histology of Neuron & Its Composition. *Journal of Neurology and Neurophysiology*, 12, 1-2. https://doi.org/10.35248/2155-9562.20.12.524.

Aydoğdu, S., & Eken, E. (2023). Calculation of cerebral hemispheres volume values (grey matter, white matter and lateral ventricle) of sheep and goat: A stereological study. *Anatomia, histologia, embryologia*. https://doi.org/10.1111/ahe.12983.

Bale, T., Abel, T., Akil, H., Carlezon, W., Moghaddam, B., Nestler, E., Ressler, K., & Thompson, S. (2019). The critical importance of basic animal research for neuropsychiatric disorders. *Neuropsychopharmacology*, 44, 1349-1353. https://doi.org/10.1038/s41386-019-0405-9.

Baloyannis, S. (2018). Neurosciences in Hellenistic Alexandria. *Journal of Neurology & Stroke*. https://doi.org/10.15406/JNSK.2018.08.00320.

Bars, D., Gozariu, M., & Cadden, S. (2001). Animal models of nociception. *Pharmacological reviews*, 53 4, 597-652.

Barab'asi, D., Bianconi, G., Bullmore, E., Burgess, M., Chung, S., Eliassi-Rad, T., George, D., Kov'acs, I., Makse, H., Papadimitriou, C., Nichols, T., Sporns, O., Stachenfeld, K., Toroczkai, Z., Towlson, E., Zador, A., Zeng, H., Barab'asi, A., Bernard, A., & Buzs'aki, G. (2023). Neuroscience Needs Network Science. *The Journal of Neuroscience*, 43, 5989 - 5995. https://doi.org/10.1523/JNEUROSCI.1014-23.2023.

Baratta, A., Brandner, A., Plasil, S., Rice, R., & Farris, S. (2022). Advancements in Genomic and Behavioral Neuroscience Analysis for the Study of Normal and Pathological Brain Function. *Frontiers in Molecular Neuroscience*, 15. https://doi.org/10.3389/fnmol.2022.905328.

Barinaga, M. (1990). Technical advances power neuroscience. *Science*, 250 4983, 908-9. https://doi.org/10.1126/SCIENCE.2237438.

Barron, H., Mars, R., Dupret, D., Lerch, J., & Sampaio-Baptista, C. (2020). Cross-species neuroscience: closing the explanatory gap. *Philosophical Transactions of the Royal Society B: Biological Sciences*, 376. https://doi.org/10.1098/rstb.2019.0633.

Barrot, M. (2012). Tests and models of nociception and pain in rodents. *Neuroscience*, 211, 39-50. https://doi.org/10.1016/j.neuroscience.2011.12.041.

Bartkowska, K., Tepper, B., Turlejski, K., & Djavadian, R. (2022). Postnatal and Adult Neurogenesis in Mammals, Including Marsupials. *Cells*, 11. https://doi.org/10.3390/cells11172735.

Bassett, D., & Sporns, O. (2017). Network neuroscience. *Nature Neuroscience*, 20, 353-364. https://doi.org/10.1038/nn.4502.

Bates, A., Manton, J., Jagannathan, S., Costa, M., Schlegel, P., Rohlfing, T., & Jefferis, G. (2020). The natverse, a versatile toolbox for combining and analysing neuroanatomical data. *eLife*, 9. https://doi.org/10.7554/eLife.53350.

Beissner, F., Meissner, K., Bär, K., & Napadow, V. (2013). The Autonomic Brain: An Activation Likelihood Estimation Meta-Analysis for Central Processing of Autonomic Function. *The Journal of Neuroscience*, 33, 10503 - 10511. https://doi.org/10.1523/JNEUROSCI.1103-13.2013.

Bell A. (2018). The neurobiology of acute pain. Veterinary journal (London, England: 1997), 237, 55–62. https://doi.org/10.1016/j.tvjl.2018.05.004.

Beniaguev, D., Segev, I., & London, M. (2019). Single cortical neurons as deep artificial neural networks. *Neuron*, 109, 2727-2739.e3. https://doi.org/10.1101/613141.

Beran, M. (2020). Animal Learning and Cognition. *Oxford Research Encyclopedia of Psychology*. https://doi.org/10.1093/acrefore/9780190236557.013.644.

Berg, D., Su, Y., Jimenez-Cyrus, D., Patel, A., Huang, N., Morizet, D., Lee, S., Shah, R., Ringeling, F., Jain, R., Epstein, J., Wu, Q., Canzar, S., Ming, G., Song, H., & Bond, A. (2019). A Common Embryonic Origin of Stem Cells Drives Developmental and Adult Neurogenesis. *Cell*, 177, 654-668.e15. https://doi.org/10.1016/j.cell.2019.02.010.

Bickle, J. (2016). Revolutions in Neuroscience: Tool Development. *Frontiers in Systems Neuroscience*, 10. https://doi.org/10.3389/fnsys.2016.00024.

Bisiacchi, P., & Cainelli, E. (2021). Structural and functional brain asymmetries in the early phases of life: a scoping review. *Brain Structure & Function*, 227, 479 - 496. https://doi.org/10.1007/s00429-021-02256-1.

Boehm, I., Alhindi, A., Leite, A., Logie, C., Gibbs, A., Murray, O., Farrukh, R., Pirie, R., Proudfoot, C., Clutton, R., Wishart, T., Jones, R., & Gillingwater, T. (2020). Comparative anatomy of the mammalian neuromuscular junction. *Journal of Anatomy*, 237, 827 - 836. https://doi.org/10.1111/joa.13260.

Bota, M., Talpalaru, S., Hintiryan, H., Dong, H., & Swanson, L. (2014). BAMS2 workspace: A comprehensive and versatile neuroinformatic platform for collating and processing neuroanatomical connections. *Journal of Comparative Neurology*, 522. https://doi.org/10.1002/cne.23592.

Blázquez-Llorca, L., De Paz, L., Martín-Orti, R., Santos-Álvarez, I., Fernández-Valle, M., Castejón, D., García-Real, I., Salgüero-Fernández, R., Pérez-Lloret, P., Moreno, N., Jiménez, S., Herrero-Fernández, M., & González-Soriano, J. (2023). The Application of 3D Anatomy for Teaching Veterinary Clinical Neurology. *Animals: an Open Access Journal from MDPI*, 13. https://doi.org/10.3390/ani13101601.

Bräuer, J., Hanuš, D., Pika, S., Gray, R., & Uomini, N. (2020). Old and New Approaches to Animal Cognition: There Is Not "One Cognition." *Journal of Intelligence*, 8. https://doi.org/10.3390/jintelligence8030028.

Brittin, C., Cook, S., Hall, D., Emmons, S., & Cohen, N. (2021). A multi-scale brain map derived from whole-brain volumetric reconstructions. *Nature*, 591, 105 - 110. https://doi.org/10.1038/s41586-021-03284-x.

Brawanski, A. (2012). On the myth of the Edwin Smith papyrus: is it magic or science?. *Acta Neurochirurgica*, 154, 2285-2291. https://doi.org/10.1007/s00701-012-1523-x.

Breitenfeld, T., Jurašić, M., & Breitenfeld, D. (2014). Hippocrates: the forefather of neurology. *Neurological Sciences*, 35, 1349-1352. https://doi.org/10.1007/s10072-014-1869-3.

Brenowitz, E., & Zakon, H. (2015). Emerging from the bottleneck: benefits of the comparative approach to modern neuroscience. *Trends in Neurosciences*, 38, 273-278. https://doi.org/10.1016/j.tins.2015.02.008.

Brown, R. (2019). Why Study the History of Neuroscience?. *Frontiers in Behavioral Neuroscience*, 13. https://doi.org/10.3389/fnbeh.2019.00082.

Buchsbaum, I., & Cappello, S. (2019). Neuronal migration in the CNS during development and disease: insights from *in vivo* and *in vitro* models. *Development*, 146. https://doi.org/10.1242/dev.163766.

Bullock, T., Bennett, M., Johnston, D., Josephson, R., Marder, E., & Fields, R. (2005). The Neuron Doctrine, Redux. *Science*, 310, 791 - 793. https://doi.org/10.1126/SCIENCE.1114394.

Bunford, N., Andics, A., Kis, A., Miklósi, Á., & Gácsi, M. (2017). Canis familiaris As a Model for Non-Invasive Comparative Neuroscience. *Trends in Neurosciences*, 40, 438-452. https://doi.org/10.1016/j.tins.2017.05.003.

Burmeister, S., & Liu, Y. (2020). Integrative Comparative Cognition: Can Neurobiology and Neurogenomics Inform Comparative Analyses of Cognitive Phenotype? *Integrative and comparative biology*, 60 4, 925-928. https://doi.org/10.1093/icb/icaa113.

Burrell, B. (2017). Comparative biology of pain: What invertebrates can tell us about how nociception works. *Journal of neurophysiology*, 117 4, 1461-1473. https://doi.org/10.1152/jn.00600.2016.

Caicoya, A., Amici, F., Ensenyat, C., & Colell, M. (2021). Comparative cognition in three understudied ungulate species: European bison, forest buffalos and giraffes. *Frontiers in Zoology*, 18. https://doi.org/10.1186/s12983-021-00417-w.

Cambiaghi, M. (2017). Andreas Vesalius (1514–1564). *Journal of Neurology*, 264, 1828-1830. https://doi.org/10.1007/s00415-017-8459-2.

Cambiaghi, M., & Hausse, H. (2019). Leonardo da Vinci (1452–1519) and the legacy of a "Renaissance neurologist." *Neurology*, 93, 717 - 718. https://doi.org/10.1212/WNL.0000000000008333.

Cao, H., Hong, X., Tost, H., Meyer-Lindenberg, A., & Schwarz, E. (2022). Advancing translational research in neuroscience through multi-task learning. *Frontiers in Psychiatry*, 13. https://doi.org/10.3389/fpsyt.2022.993289.

Carlson, B. M. (2019). *The Human Body || The Nervous System*. 137–175. https://doi.org/10.1016/B978-0-12-804254-0.00006-5.

Carson, R. (2020). Inter-hemispheric inhibition sculpts the output of neural circuits by co-opting the two cerebral hemispheres. *The Journal of Physiology*, 598. https://doi.org/10.1113/JP279793.

Carr, F., & Zachariou, V. (2014). Nociception and pain: lessons from optogenetics. *Frontiers in Behavioral Neuroscience*, 8. https://doi.org/10.3389/fnbeh.2014.00069.

Casey, K. (1999). Forebrain mechanisms of nociception and pain: analysis through imaging. *Proceedings of the National Academy of Sciences of the United States of America*, 96 14, 7668-74. https://doi.org/10.1073/PNAS.96.14.7668.

Celsus A (1465) De medicina libri VIII. Anonymous Handwriter, Florence.

Chang, A., Lad, E., & Lad, S. (2007). Hippocrates' influence on the origins of neurosurgery. *Neurosurgical focus*, 23 1, E9. https://doi.org/10.3171/FOC-07/07/E9.

Chen, X., Zhan, H., Kebschull, J., Sun, Y., & Zador, A. (2019). High-Throughput Mapping of Long-Range Neuronal Projection Using In Situ Sequencing. *Cell*, 179, 772-786.e19. https://doi.org/10.1016/j.cell.2019.09.023.

Chrysostomou, E., Flici, H., Gornik, S., Salinas-Saavedra, M., Gahan, J., McMahon, E., Thompson, K., Hanley, S., Kilcoyne, M., Schnitzler, C., Gonzalez, P., Baxevanis, A., & Frank, U. (2022). A cellular and molecular analysis of SoxB-driven neurogenesis in a cnidarian. *eLife*, 11. https://doi.org/10.1101/2022.03.22.485278.

Chung, M., & Chung, B. (2020). *Morphology of the central nervous system*, 1-43. https://doi.org/10.1016/b978-0-12-819901-5.00001-6.

Chvátal, A. (2015). Discovering the Structure of Nerve Tissue: Part 2: Gabriel Valentin, Robert Remak, and Jan Evangelista Purkyně. *Journal of the History of the Neurosciences*, 24, 326 - 351. https://doi.org/10.1080/0964704X.2014.977677.

Cicchetti, D., & Dawson, G. (2002). Editorial: Multiple levels of analysis. *Development and Psychopathology*, 14, 417 - 420. https://doi.org/10.1017/S0954579402003012.

Cowan, M., Harter, D., & Kandel, E. (2000). The emergence of modern neuroscience: some implications for neurology and psychiatry. *Annual review of neuroscience*, 23, 343-91. https://doi.org/10.1146/annurev.neuro.23.1.343.

Cozzi, B., Bonfanti, L., Canali, E., & Minero, M. (2020). Brain Waste: The Neglect of Animal Brains. *Frontiers in Neuroanatomy*, 14. https://doi.org/10.3389/fnana.2020.573934.

Crawford, L., & Caterina, M. (2020). Functional Anatomy of the Sensory Nervous System: Updates From the Neuroscience Bench. *Toxicologic Pathology*, 48, 174 - 189. https://doi.org/10.1177/0192623319869011.

D'Elia, A., Schiavi, S., Soluri, A., Massari, R., Soluri, A., & Trezza, V. (2020). Role of Nuclear Imaging to Understand the Neural Substrates of Brain Disorders in Laboratory Animals: Current Status and Future Prospects. *Frontiers in Behavioral Neuroscience*, 14. https://doi.org/10.3389/fnbeh.2020.596509.

Dan, B. (2019). Neuroscience underlying rehabilitation: what is neuroplasticity?. *Developmental Medicine & Child Neurology*, 61. https://doi.org/10.1111/dmcn.14341.

Daly, B.P. (2012). *Encyclopedia of Human Behavior || Central Nervous System*. 454–459. https://doi.org/10.1016/b978-0-12-375000-6.00084-7

Dasgupta, K., & Jeong, J. (2019). Developmental biology of the meninges. *genesis*, 57. https://doi.org/10.1002/dvg.23288.

Davatzikos, C. (2019). Machine learning in neuroimaging: Progress and challenges. *NeuroImage*, 197, 652-656. https://doi.org/10.1016/j.neuroimage.2018.10.003.

De Schotten, T., Croxson, P., & Mars, R. (2019). Large-scale comparative neuroimaging: Where are we and what do we need?. *Cortex; a Journal Devoted to the Study of the Nervous System and Behavior*, 118, 188 - 202. https://doi.org/10.1016/j.cortex.2018.11.028.

Decimo, I., Dolci, S., Panuccio, G., Riva, M., Fumagalli, G., & Bifari, F. (2020). Meninges: A Widespread Niche of Neural Progenitors for the Brain. *The Neuroscientist*, 27, 506 - 528. https://doi.org/10.1177/1073858420954826.

Deisseroth, K., & Schnitzer, M. (2013). Engineering Approaches to Illuminating Brain Structure and Dynamics. *Neuron*, 80, 568-577. https://doi.org/10.1016/j.neuron.2013.10.032.

DeNardo, L., & Luo, L. (2017). Genetic strategies to access activated neurons. *Current Opinion in Neurobiology*, 45, 121-129. https://doi.org/10.1016/j.conb.2017.05.014.

Dennis, E., Hady, A., Michaiel, A., Clemens, A., Tervo, D., Voigts, J., & Datta, S. (2020). Systems Neuroscience of Natural Behaviors in Rodents. *The Journal of Neuroscience*, 41, 911 - 919. https://doi.org/10.31219/osf.io/y5fbq.

Derk, J., Jones, H., Como, C., Pawlikowski, B., & Siegenthaler, J. (2021). Living on the Edge of the CNS: Meninges Cell Diversity in Health and Disease. *Frontiers in Cellular Neuroscience*, 15. https://doi.org/10.3389/fncel.2021.703944.

Detwiler, L., Mejino, J., & Brinkley, J. (2016). From frames to OWL2: Converting the Foundational Model of Anatomy. *Artificial intelligence in medicine*, 69, 12-21. https://doi.org/10.1016/j.artmed.2016.04.003.

Dockès, J., Poldrack, R., Primet, R., Gözükan, H., Yarkoni, T., Suchanek, F., Thirion, B., & Varoquaux, G. (2020). NeuroQuery, comprehensive meta-analysis of human brain mapping. *eLife*, 9. https://doi.org/10.7554/eLife.53385.

Domínguez-Oliva, A., Mota-Rojas, D., Hernández-Avalos, I., Mora-Medina, P., Olmos-Hernández, A., Verduzco-Mendoza, A., Casas-Alvarado, A., & Whittaker, A. L.

(2022). The neurobiology of pain and facial movements in rodents: Clinical applications and current research. *Frontiers in veterinary science*, 9, 1016720. https://doi.org/10.3389/fvets.2022.1016720.

Donkelaar, H., & Puelles, L. (2019). Editorial: Recent Developments in Neuroanatomical Terminology. *Frontiers in Neuroanatomy*, 13. https://doi.org/10.3389/fnana.2019.00080.

Donner, K., & Yovanovich, C. (2020). A frog's eye view: Foundational revelations and future promises. *Seminars in cell & developmental biology*. https://doi.org/10.1016/j.semcdb.2020.05.011.

Dubin, A., & Patapoutian, A. (2010). Nociceptors: the sensors of the pain pathway. *The Journal of clinical investigation*, 120 11, 3760-72. https://doi.org/10.1172/JCI42843.

Dufour, S., Quérat, B., Tostivint, H., Pasqualini, C., Vaudry, H., & Rousseau, K. (2020). Origin and Evolution of the Neuroendocrine Control of Reproduction in Vertebrates, with Special Focus on Genome and Gene Duplications. *Physiological reviews*. https://doi.org/10.1152/physrev.00009.2019.

Duparc, F. (2021). Visually memorable neuroanatomy for beginners. *Surgical and Radiologic Anatomy*, 43, 615. https://doi.org/10.1007/s00276-021-02728-3.

Ernst, G. (2014). *The Autonomic Nervous System*. 27-49. https://doi.org/10.1007/978-1-4471-4309-3_3.

El-Sayes, J., Harasym, D., Turco, C. V., Locke, M. B., & Nelson, A. J. (2019). Exercise-Induced Neuroplasticity: A Mechanistic Model and Prospects for Promoting Plasticity. *The Neuroscientist: a review journal bringing neurobiology, neurology and psychiatry*, 25(1), 65–85. https://doi.org/10.1177/1073858418771538.

Favela, L. (2020). The dynamical renaissance in neuroscience. *Synthese*, 199, 2103-2127. https://doi.org/10.1007/S11229-020-02874-Y.

Feldman, R., & Goodrich, J. (1999). The Edwin Smith Surgical Papyrus. *Child's Nervous System*, 15, 281-284. https://doi.org/10.1007/s003810050395.

Ferraro, S., Klugah-Brown, B., Tench, C., Bore, M., Nigri, A., Demichelis, G., Bruzzone, M., Palermo, S., Zhao, W., Yao, S., Jiang, X., Kendrick, K., & Becker, B. (2022). The central autonomic system revisited – Convergent evidence for a regulatory role of the insular and midcingulate cortex from neuroimaging meta-analyses. *Neuroscience & Biobehavioral Reviews*, 142. https://doi.org/10.1101/2022.05.25.493371.

Finger S, Fernando HR. E. George Squier and the Discovery of Cranial Trepanation: A Landmark in the History of Surgery and Ancient Medicine. *J of History of Medicine* 2001;56:353-381.

Flemming, K. (2021). Cortex: Gross Anatomy. *Mayo Clinic Neurology Board Review*. https://doi.org/10.1093/med/9780197512166.003.0020.

Fodstad, H. (2002). The Neuron Theory. *Stereotactic and Functional Neurosurgery*, 77, 20 - 24. https://doi.org/10.1159/000064596.

Fodstad, H., Kondziolka, D., & De Lotbinière, A. (2000). The Neuron Doctrine, the Mind, and the Arctic. *Neurosurgery*, 47, 1381-1389. https://doi.org/10.1097/00006123-200107000-00053.

Friedrich, P., Forkel, S., Amiez, C., Balsters, J., Coulon, O., Fan, L., Goulas, A., Hadj-Bouziane, F., Hecht, E., Heuer, K., Jiang, T., Latzman, R., Liu, X., Loh, K., Patil, K., Lopez-Persem, A., Procyk, E., Sallet, J., Toro, R., Vickery, S., Weis, S., Wilson, C.,

Xu, T., Zerbi, V., Eickhoff, S., Margulies, D., Mars, R., & Schotten, M. (2020). Imaging evolution of the primate brain: the next frontier?. *NeuroImage*, 228, 117685 - 117685. https://doi.org/10.1016/j.neuroimage.2020.117685.

Friedman, A., & Polson, M. (1981). Hemispheres as independent resource systems: limited-capacity processing and cerebral specialization. *Journal of Experimental Psychology: Human Perception and Performance.* https://doi.org/10.1037/0096-1523.7.5.1031.

Gage, F. (2015). Neuroscience: The Study of the Nervous System & Its Functions. *Daedalus*, 144, 5-9. https://doi.org/10.1162/DAED_e_00313.

Gage, F. (2019). Adult neurogenesis in mammals. *Science*, 364, 827 - 828. https://doi.org/10.1126/science.aav6885.

Gallant, J., & O'Connell, L. (2020). Studying convergent evolution to relate genotype to behavioral phenotype. *Journal of Experimental Biology*, 223. https://doi.org/10.1242/jeb.213447.

Galsworthy, M., Arden, R., & Chabris, C. (2014). *Animal Models of General Cognitive Ability for Genetic Research into Cognitive Functioning*. 257-278. https://doi.org/10.1007/978-1-4614-7447-0_9.

Gao, J., Yang, C., Li, Q., Chen, L., Jiang, Y., Liu, S., Zhang, J., Liu, G., & Chen, J. (2020). Hemispheric Difference of Regional Brain Function Exists in Patients With Acute Stroke in Different Cerebral Hemispheres: A Resting-State fMRI Study. *Frontiers in Aging Neuroscience*, 13. https://doi.org/10.3389/fnagi.2021.691518.

Gao, P., & Ganguli, S. (2015). On simplicity and complexity in the brave new world of large-scale neuroscience. *Current Opinion in Neurobiology*, 32, 148-155. https://doi.org/10.1016/j.conb.2015.04.003.

Gato, A., Alonso, M., Lamus, F., & Miyan, J. (2019). Neurogenesis: A process ontogenically linked to brain cavities and their content, CSF. *Seminars in cell & developmental biology*. https://doi.org/10.1016/j.semcdb.2019.11.008.

Gerlai, R. (2020). Evolutionary conservation, translational relevance and cognitive function: The future of zebrafish in behavioral neuroscience. *Neuroscience & Biobehavioral Reviews*, 116, 426-435. https://doi.org/10.1016/j.neubiorev.2020.07.009.

Geschwind, D., & Konopka, G. (2009). Neuroscience in the era of functional genomics and systems biology. *Nature*, 461, 908-915. https://doi.org/10.1038/nature08537.

Ghosh, S., & Narayan, R. (2020). Anatomy of nervous system and emergence of neuroscience: A chronological journey across centuries. *Morphologie: bulletin de l'Association des anatomistes*. https://doi.org/10.1016/j.morpho.2020.05.005.

Gibbons, C. (2019). Basics of autonomic nervous system function. Handbook of clinical neurology, 160, 407-418. https://doi.org/10.1016/B978-0-444-64032-1.00027-8.

Gillespie, T., Tripathy, S., Sy, M., Martone, M., & Hill, S. (2020). The Neuron Phenotype Ontology: A FAIR Approach to Proposing and Classifying Neuronal Types. *Neuroinformatics*, 20, 793 - 809. https://doi.org/10.1007/s12021-022-09566-7.

Goodwin, R., Black, A., & Nathaniel, T. (2023). Integrating basic, clinical, and health system science in a medical neuroscience course of an integrated pre-clerkship curriculum. *Anatomical sciences education*. https://doi.org/10.1002/ase.2343.

Goralskyi, L., Sokulskyi, I., Kolesnik, N., Dunaievska, O., Radzykhovskyi, N., Gutyj, B., & Shevchuk, S. (2023). Evolutionary morphology of spinal nodes of poikilotherm vertebrate animals. *Scientific Messenger of LNU of Veterinary Medicine and Biotechnologies*. https://doi.org/10.32718/nvlvet10908.

Gordleeva, S., Tsybina, Y., Krivonosov, M., Tyukin, I., Kazantsev, V., Zaikin, A., & Gorban, A. (2023). Situation-Based Neuromorphic Memory in Spiking Neuron-Astrocyte Network. *IEEE transactions on neural networks and learning systems*, PP. https://doi.org/10.1109/TNNLS.2023.3335450.

Gross, C. (2009). *History of Neuroscience: Early Neuroscience*. 1167-1171. https://doi.org/10.1016/B978-008045046-9.00992-X.

Guerrini, A. (2018). Human & Animal Cognition in Early Modern Philosophy & Medicine ed. by Stefanie Buchenau and Roberto Lo Presti (review). *Bulletin of the History of Medicine*, 92, 377 - 378. https://doi.org/10.1353/bhm.2018.0035.

Guidolin, D., Tortorella, C., Marcoli, M., Maura, G., & Agnati, L. (2022). Intercellular Communication in the Central Nervous System as Deduced by Chemical Neuroanatomy and Quantitative Analysis of Images: Impact on Neuropharmacology. *International Journal of Molecular Sciences*, 23. https://doi.org/10.3390/ijms 23105805.

Guillery, R. (2005). Observations of synaptic structures: origins of the neuron doctrine and its current status. *Philosophical Transactions of the Royal Society B: Biological Sciences*, 360, 1281 - 1307. https://doi.org/10.1098/rstb.2003.1459.

Gupta, P., Sharma, A., Gundamaraju, R., Pant, K., Mani, S., Singh, S., Maccioni, R., Datta, T., Singh, S., Gupta, G., Prasher, P., Dua, K., Dey, A., Sharma, C., Mughal, Y., Ruokolainen, J., Kesari, K., & Ojha, S. (2022). Molecular mechanisms of developmental pathways in neurological disorders: a pharmacological and therapeutic review. *Open Biology*, 12. https://doi.org/10.1098/rsob.210289.

Gurule, S., Sustaita-Monroe, J., Padmanabhan, V., & Cardoso, R. (2023). Developmental programming of the neuroendocrine axis by steroid hormones: Insights from the sheep model of PCOS. *Frontiers in Endocrinology*, 14. https://doi.org/10.3389/fendo.2023.1096187.

Hale, M. (2014). Mapping Circuits beyond the Models: Integrating Connectomics and Comparative Neuroscience. *Neuron*, 83, 1256-1258. https://doi.org/10.1016/j.neuron.2014.08.032.

Hart, J. (2020). Basic neuroanatomy review. *Oxford Textbook of Neuropsychiatry*. https://doi.org/10.1093/med/9780198757139.003.0006.

Hassabis, D., Kumaran, D., Summerfield, C., & Botvinick, M. (2017). Neuroscience-Inspired Artificial Intelligence. *Neuron*, 95, 245-258. https://doi.org/10.1016/j.neuron.2017.06.011.

He, J., Li, B., Han, S., Zhang, Y., Liu, K., Yi, S., Liu, Y., & Xiu, M. (2022). Drosophila as a Model to Study the Mechanism of Nociception. *Frontiers in Physiology*, 13. https://doi.org/10.3389/fphys.2022.854124.

Healy, S. (2019). The face of animal cognition. *Integrative zoology*, 14 2, 132-144. https://doi.org/10.1111/1749-4877.12361.

Hentzen, N., Ferretti, M., Chadha, S., Jaarsma, J., De Visser, M., & Moro, E. (2022). Mapping of European activities on the integration of sex and gender factors in neurology and neuroscience. *European Journal of Neurology*, 29, 2572 - 2579. https://doi.org/10.1111/ene.15439.

Herculano-Houzel, S., Catania, K., Manger, P., & Kaas, J. (2015). Mammalian Brains Are Made of These: A Dataset of the Numbers and Densities of Neuronal and Nonneuronal Cells in the Brain of Glires, Primates, Scandentia, Eulipotyphlans, Afrotherians and Artiodactyls, and Their Relationship with Body Mass. *Brain, Behavior and Evolution*, 86, 145 - 163. https://doi.org/10.1159/000437413.

Hines, M., & Carnevale, N. (2001). Neuron: A Tool for Neuroscientists. *The Neuroscientist*, 7, 123 - 135. https://doi.org/10.1177/107385840100700207.

Homberg, J., Adan, R., Alenina, N., Asiminas, A., Bader, M., Beckers, T., Begg, D., Blokland, A., Burger, M., Dijk, G., Eisel, U., Elgersma, Y., Englitz, B., Fernández-Ruiz, A., Fitzsimons, C., Dam, A., Gass, P., Grandjean, J., Havekes, R., Henckens, M., Herden, C., Hut, R., Jarrett, W., Jeffrey, K., Jezova, D., Kalsbeek, A., Kamermans, M., Kas, M., Kasri, N., Kiliaan, A., Kolk, S., Korosi, A., Korte, S., Kozicz, T., Kushner, S., Leech, K., Lesch, K., Lesscher, H., Lucassen, P., Luthi, A., , L., Mallien, A., Meerlo, P., Mejías, J., Meye, F., Mitchell, A., Mul, J., Olcese, U., González, A., Olivier, J., Pasqualetti, M., Pennartz, C., Popik, P., Prickaerts, J., Prida, L., Ribeiro, S., Roozendaal, B., Rossato, J., Salari, A., Schoemaker, R., Smit, A., Vanderschuren, L., Takeuchi, T., Veen, R., Smidt, M., Vyazovskiy, V., Wiesmann, M., Wierenga, C., Williams, B., Willuhn, I., Wöhr, M., Wolvekamp, M., Zee, E., & Genzel, L. (2021). The continued need for animals to advance brain research. *Neuron*, 109, 2374-2379. https://doi.org/10.1016/j.neuron.2021.07.015.

Holguera, I., & Desplan, C. (2018). Neuronal specification in space and time. *Science*, 362, 176 - 180. https://doi.org/10.1126/science.aas9435.

Hughes, J. (1988). The Edwin Smith Surgical Papyrus: an analysis of the first case reports of spinal cord injuries. *Paraplegia*, 26, 71-82. https://doi.org/10.1038/sc.1988.15.

Huang, Z., Paul, A., & Paul, A. (2019). The diversity of GABAergic neurons and neural communication elements. *Nature Reviews Neuroscience*, 20, 563 - 572. https://doi.org/10.1038/s41583-019-0195-4.

Hückesfeld, S., Schlegel, P., Miroschnikow, A., Schoofs, A., Zinke, I., Haubrich, A., Schneider-Mizell, C., Truman, J., Fetter, R., Cardona, A., & Pankratz, M. (2020). Unveiling the sensory and interneuronal pathways of the neuroendocrine system in Drosophila. *bioRxiv*. https://doi.org/10.1101/2020.10.22.350306.

Hückesfeld, S., Schlegel, P., Miroschnikow, A., Schoofs, A., Zinke, I., Haubrich, A., Schneider-Mizell, C., Truman, J., Fetter, R., Cardona, A., & Pankratz, M. (2021). Unveiling the sensory and interneuronal pathways of the neuroendocrine connectome in Drosophila. *eLife*, 10. https://doi.org/10.7554/eLife.65745.

Ikkert, O., & Korol, T. (2023). Cerebral Hemispheres Functional Asymmetry in the Education of Natural Specialties And Humanities Students. *Scientific Journal of Polonia University*. https://doi.org/10.23856/5813.

Ioanas, H., Schlegel, F., Skachokova, Z., Schroeter, A., Husak, T., & Rudin, M. (2022). Hybrid fiber optic-fMRI for multimodal cell-specific recording and manipulation of neural activity in rodents. *Neurophotonics*, 9. https://doi.org/10.1117/1.NPh.9.3.032206.

Irwin, L., & Irwin, B. (2020). Place and Environment in the Ongoing Evolution of Cognitive Neuroscience. *Journal of Cognitive Neuroscience*, 32, 1837-1850. https://doi.org/10.1162/jocn_a_01607.

Jänig, W. (2022). *The Integrative Action of the Autonomic Nervous System.* https://doi.org/10.1017/CBO9780511541667.017.

Janmaat, K., De Guinea, M., Collet, J., Byrne, R., Robira, B., Van Loon, E., Jang, H., Biro, D., Ramos-Fernández, G., Ross, C., Presotto, A., Allritz, M., Alavi, S., & Van Belle, S. (2021). Using natural travel paths to infer and compare primate cognition in the wild. *iScience*, 24. https://doi.org/10.1016/j.isci.2021.102343.

Jha, N., Chen, W., Kumar, S., Dubey, R., Tsai, L., Kar, R., Jha, S., Bodmer, B., Mückschel, M., Roessner, V., & Beste, C. (2017). Neurophysiological variability masks differences in functional neuroanatomical networks and their effectiveness to modulate response inhibition between children and adults. *Brain Structure and Function*, 223, 1797 - 1810. https://doi.org/10.1007/s00429-017-1589-6.

Julius, D. (2013). TRP channels and pain. *Annual review of cell and developmental biology*, 29, 355-84. https://doi.org/10.1146/annurev-cellbio-101011-155833.

Jimsheleishvili S, Dididze M. Neuroanatomy, Cerebellum. [Updated 2023 Jul 24]. In: *StatPearls* [Internet]. Treasure Island (FL): StatPearls Publishing; 2024 Jan-. Available from: https://www.ncbi.nlm.nih.gov/books/NBK538167/.

Kai, W. (2002). The Development of Neuroscience. *Medicine and Philosophy*.

Karemaker, J. (2017). An introduction into autonomic nervous function. *Physiological Measurement*, 38, R89 - R118. https://doi.org/10.1088/1361-6579/aa6782.

Kang, K., Pulver, S. R., Panzano, V. C., Chang, E. C., Griffith, L. C., Theobald, D. L., & Garrity, P. A. (2010). Analysis of Drosophila TRPA1 reveals an ancient origin for human chemical nociception. *Nature*, 464(7288), 597–600. https://doi.org/10.1038/nature08848

Karenberg A (2005) Amor, Äskulap & Co. Klassische Mythologie in der Sprache der modernen Medizin. Schattauer, Stuttgart.

Kashetsky, T., Avgar, T., & Dukas, R. (2021). *The Cognitive Ecology of Animal Movement: Evidence From Birds and Mammals*. 9. https://doi.org/10.3389/fevo.2021.724887.

Keifer, J., & Summers, C. (2016). Putting the "Biology" Back into "Neurobiology": The Strength of Diversity in Animal Model Systems for Neuroscience Research. *Frontiers in Systems Neuroscience*, 10. https://doi.org/10.3389/fnsys.2016.00069.

Kell, A., & McDermott, J. (2019). Deep neural network models of sensory systems: windows onto the role of task constraints. *Current Opinion in Neurobiology*, 55, 121-132. https://doi.org/10.1016/j.conb.2019.02.003.

Kelly, A., & Wilson, L. (2020). Aggression: Perspectives from social and systems neuroscience. *Hormones and Behavior*, 123. https://doi.org/10.1016/j.yhbeh.2019.04.010.

Khiroug, L., & Verkhratsky, A. (2021). Coming full circle: In vivo Veritas, or expanding the neuroscience frontier. *Cell calcium*, 98, 102452. https://doi.org/10.1016/j.ceca.2021.102452.

Khonsary, S. (2017). THE BRAIN, An Introduction to Functional Neuroanatomy. *Surgical Neurology International*, 8. https://doi.org/10.4103/SNI.SNI_47_17.

Krichbaum, S., Davila, A., Lazarowski, L., & Katz, J. (2020). Animal Cognition. *Oxford Research Encyclopedia of Psychology*. https://doi.org/10.1093/acrefore/9780190236557.013.648.

Kriegeskorte, N., & Douglas, P. (2018). Cognitive computational neuroscience. *Nature Neuroscience*, 21, 1148 - 1160. https://doi.org/10.1038/s41593-018-0210-5.

La Rosa, C., & Bonfanti, L. (2018). Brain Plasticity in Mammals: An Example for the Role of Comparative Medicine in the Neurosciences. *Frontiers in Veterinary Science*, 5. https://doi.org/10.3389/fvets.2018.00274.

LaBar, K. (2017). Advances in neuroscience. *Science Advances*, 3. https://doi.org/10.1126/sciadv.aar2953.

Lambert, K. (2005). The Clinical Neuroscience Course: Viewing Mental Health from Neurobiological Perspectives. *Journal of Undergraduate Neuroscience Education*, 3, A42 - A52.

Lanciego, J., & Wouterlood, F. (2020). Neuroanatomical tract-tracing techniques that did go viral. *Brain Structure & Function*, 225, 1193 - 1224. https://doi.org/10.1007/s00429-020-02041-6.

Lanigan, L., Russell, D., Woolard, K., Pardo, I., Godfrey, V., Jortner, B., Butt, M., & Bolon, B. (2020). Comparative Pathology of the Peripheral Nervous System. *Veterinary Pathology*, 58, 10 - 33. https://doi.org/10.1177/0300985820959231.

Lee, S. (2019). The History of Neuroscience 1: Ancient Neuroscience. *Epilia: Epilepsy and Community*. https://doi.org/10.35615/epilia.2019.00010.

Lee, G., Jang, H., & Oh, Y. (2023). The role of diuretic hormones (DHs) and their receptors in Drosophila. *BMB Reports*, 56, 209 - 215. https://doi.org/10.5483/BMBRep.2023-0021.

Leng, G., & MacGregor, D. (2018). Models in neuroendocrinology. *Mathematical biosciences*, 305, 29-41. https://doi.org/10.1016/j.mbs.2018.07.008.

Lerch, J., Van Der Kouwe, A., Raznahan, A., Paus, T., Johansen-Berg, H., Miller, K., Smith, S., Fischl, B., & Sotiropoulos, S. (2017). Studying neuroanatomy using MRI. *Nature Neuroscience*, 20, 314-326. https://doi.org/10.1038/nn.4501.

Leung, E., & Roberto, A. (2017). Behavioral Neuroscience 8th Edition. *The Yale Journal of Biology and Medicine*, 90, 157 - 158.

Lewis, A., Calipari, E., & Siciliano, C. (2021). Toward Standardized Guidelines for Investigating Neural Circuit Control of Behavior in Animal Research. *eNeuro*, 8. https://doi.org/10.1523/ENEURO.0498-20.2021.

Lim, L., Mi, D., Llorca, A., & Marín, O. (2018). Development and Functional Diversification of Cortical Interneurons. *Neuron*, 100, 294-313. https://doi.org/10.1016/j.neuron.2018.10.009.

Lim, Y., & Golden, J. A. (2007). Patterning the developing diencephalon. *Brain research reviews*, 53(1), 17–26. https://doi.org/10.1016/j.brainresrev.2006.06.004.

Lima, A., Mridha, F., Das, S., Kabir, M., Islam, M., & Watanobe, Y. (2022). A Comprehensive Survey on the Detection, Classification, and Challenges of Neurological Disorders. *Biology*, 11. https://doi.org/10.3390/biology11030469.

Lin, X., Duan, X., Jacobs, C., Ullmann, J., Chan, C., Chen, S., Cheng, S., Zhao, W., Poduri, A., Wang, X., Haggarty, S., & Shi, P. (2018). High-throughput brain activity mapping and machine learning as a foundation for systems neuropharmacology. *Nature Communications*, 9. https://doi.org/10.1038/s41467-018-07289-5.

Lisman, J. (2015). The Challenge of Understanding the Brain: Where We Stand in 2015. *Neuron*, 86, 864-882. https://doi.org/10.1016/j.neuron.2015.03.032.

Litwińczuk, M., Trujillo-Barreto, N., Muhlert, N., Cloutman, L., & Woollams, A. (2022). Relating Cognition to both Brain Structure and Function: A Systematic Review of Methods. *Brain connectivity*. https://doi.org/10.1089/brain.2022.0036.

Liu, Y., & Konopka, G. (2020). An integrative understanding of comparative cognition: lessons from human brain evolution. *Integrative and comparative biology*. https://doi.org/10.1093/icb/icaa109.

Luo, L. (2021). Architectures of neuronal circuits. *Science*, 373. https://doi.org/10.1126/science.abg7285.

Macrì, S., & Di-Poï, N. (2020). Heterochronic Developmental Shifts Underlying Squamate Cerebellar Diversity Unveil the Key Features of Amniote Cerebellogenesis. *Frontiers in Cell and Developmental Biology*, 8. https://doi.org/10.3389/fcell.2020.593377.

Maden, M., Serrano, N., Bermudez, M., & Sandoval, A. (2020). A profusion of neural stem cells in the brain of the spiny mouse, Acomys cahirinus. *Journal of Anatomy*, 238. https://doi.org/10.1111/joa.13373.

Magoski, N. (2017). *Electrical Synapses and Neuroendocrine Cell Function*. 137-160. https://doi.org/10.1016/B978-0-12-803471-2.00007-2.

Malafoglia, V., Bryant, B., Raffaeli, W., Giordano, A., & Bellipanni, G. (2013). The zebrafish as a model for nociception studies. *Journal of Cellular Physiology*, 228. https://doi.org/10.1002/jcp.24379.

Mandino, F., Vujic, S., Grandjean, J., & Lake, E. (2023). Where do we stand on fMRI in awake mice?. *Cerebral cortex*. https://doi.org/10.1093/cercor/bhad478.

Marais, O., Tourigny, K., Lellis-Santos, C., Burke, K., Pucci, N., Riisager, O., Jeon, S., Zhang, W., Lastovikova, J., Kopak, J., & Krebs, C. (2020). Cerebro, the Virtual Lab - a mobile neuroanatomy game that is both fun and educational. *The FASEB Journal*, 34. https://doi.org/10.1096/fasebj.2020.34.s1.06821.

Mars, R., Neubert, F., Verhagen, L., Sallet, J., Miller, K., Dunbar, R., & Barton, R. (2014). Primate comparative neuroscience using magnetic resonance imaging: promises and challenges. *Frontiers in Neuroscience*, 8. https://doi.org/10.3389/fnins.2014.00298.

Mars, R., Jbabdi, S., & Rushworth, M. (2021). A Common Space Approach to Comparative Neuroscience. *Annual review of neuroscience*. https://doi.org/10.1146/annurev-neuro-100220-025942.

Martín-Araguz, A., Bustamante-Martínez, C., Ajo, F., & Moreno-Martinez, J. (2002). [Neuroscience in Al Andalus and its influence on medieval scholastic medicine]. *Revista de neurologia*, 34 9, 877-92. https://doi.org/10.33588/RN.3409.2001382.

Maryenko, N., & Stepanenko, O. (2023). Spatial and Structural Complexity of Cerebral Hemispheres in Male and Female Brain: Fractal and Quantitative Analyses of MRI Brain Scans. *Revista Argentina de Anatomía Clínica*. https://doi.org/10.31051/1852.8023.v15.n3.43151.

Mateos-Aparicio, P., & Rodríguez-Moreno, A. (2019). The Impact of Studying Brain Plasticity. *Frontiers in Cellular Neuroscience*, 13. https://doi.org/10.3389/fncel.2019.00066.

Mathuru, A., Libersat, F., Vyas, A., & Teseo, S. (2020). Why behavioral neuroscience still needs diversity?: A curious case of a persistent need. *Neuroscience & Biobehavioral Reviews*, 116, 130-141. https://doi.org/10.1016/j.neubiorev.2020.06.021.

McCleskey, E., & Gold, M. (2020). Ion channels of nociception. *Annual review of physiology*, 61, 835-56. https://doi.org/10.1146/ANNUREV.PHYSIOL.61.1.835.

Millan, M. (1999). The induction of pain: an integrative review. *Progress in Neurobiology*, 57, 1-164. https://doi.org/10.1016/S0301-0082(98)00048-3.

Miller, C., Hale, M., Okano, H., Okabe, S., & Mitra, P. (2019). Comparative Principles for Next-Generation Neuroscience. *Frontiers in Behavioral Neuroscience*, 13. https://doi.org/10.3389/fnbeh.2019.00012.

Miller, J., Voorhies, W., Lurie, D., D'Esposito, M., & Weiner, K. (2020). Overlooked Tertiary Sulci Serve as a Meso-Scale Link between Microstructural and Functional Properties of Human Lateral Prefrontal Cortex. *The Journal of Neuroscience*, 41, 2229 - 2244. https://doi.org/10.1101/2020.03.24.006577.

Minagar, A., Ragheb, J., & Kelley, R. (2003). The Edwin Smith Surgical Papyrus: Description and Analysis of the Earliest Case of Aphasia. *Journal of Medical Biography*, 11, 114 - 117. https://doi.org/10.1177/096777200301100214.

Mo, A., Mukamel, E., Davis, F., Luo, C., Henry, G., Picard, S., Urich, M., Nery, J., Sejnowski, T., Lister, R., Eddy, S., Ecker, J., & Nathans, J. (2015). Epigenomic Signatures of Neuronal Diversity in the Mammalian Brain. *Neuron*, 86, 1369-1384. https://doi.org/10.1016/j.neuron.2015.05.018.

Mogil, J., & Crager, S. (2004). What should we be measuring in behavioral studies of chronic pain in animals?. *Pain*, 112, 12-15. https://doi.org/10.1016/j.pain.2004.09.028.

Mohamed, W. (2022). Arab-Muslims Contributions to Modern Neuroscience: What's Giving Us Hope?. https://doi.org/10.31730/osf.io/q4dsm.

Montgomery, A. (2013). Toward The Integration of Neuroscience and Clinical Social Work. *Journal of Social Work Practice*, 27, 333 - 339. https://doi.org/10.1080/02650533.2013.818947.

Morand-Ferron, J., Cole, E., & Quinn, J. (2016). Studying the evolutionary ecology of cognition in the wild: a review of practical and conceptual challenges. Biological Reviews, 91. https://doi.org/10.1111/brv.12174.

Moreno-Jiménez, E., Flor-García, M., Terreros-Roncal, J., Rábano, A., Cafini, F., Pallas-Bazarra, N., Ávila, J., & Llorens-Martín, M. (2019). Adult hippocampal neurogenesis is abundant in neurologically healthy subjects and drops sharply in patients with Alzheimer's disease. *Nature Medicine, 25, 554 - 560.* https://doi.org/10.1038/s41591-019-0375-9.

Moreno, R., & Holodny, A. (2021). Functional Brain Anatomy. *Neuroimaging clinics of North America*, 31 1, 33-51. https://doi.org/10.1016/j.nic.2020.09.008.

Morosanu, C., Nicolae, L., Moldovan, R., Farcasanu, A., Filip, G., & Florian, I. (2018). Neurosurgical cadaveric and in vivo large animal training models for cranial and spinal approaches and techniques - a systematic review of the current literature. *Neurologia i neurochirurgia polska*, 53 1, 8-17. https://doi.org/10.5603/PJNNS.a2019.0001.

Müller, H., Roselli, F., Rasche, V., & Kassubek, J. (2020). Diffusion Tensor Imaging-Based Studies at the Group-Level Applied to Animal Models of Neurodegenerative Diseases. *Frontiers in Neuroscience*, 14. https://doi.org/10.3389/fnins.2020.00734.

Münch, D., Ezra-Nevo, G., Francisco, A., Tastekin, I., & Ribeiro, C. (2019). Nutrient homeostasis — translating internal states to behavior. *Current Opinion in Neurobiology*, 60, 67-75. https://doi.org/10.1016/j.conb.2019.10.004.

Murtazina, A., & Adameyko, I. (2023). The peripheral nervous system. *Development*, 150 9. https://doi.org/10.1242/dev.201164.

Newhouse, A., & Chemali, Z. (2019). Neuroendocrine Disturbances in Neurodegenerative Disorders: A Scoping Review. *Psychosomatics*. https://doi.org/10.1016/j.psym. 2019.11.002.

Nicaise, A., Willis, C., Crocker, S., & Pluchino, S. (2020). Stem Cells of the Aging Brain. *Frontiers in Aging Neuroscience*, 12. https://doi.org/10.3389/fnagi.2020.00247.

Nichols, B., Mejino, J., Detwiler, L., Nilsen, T., Martone, M., Turner, J., Rubin, D., & Brinkley, J. (2014). Neuroanatomical domain of the foundational model of anatomy ontology. *Journal of Biomedical Semantics*, 5, 1 - 1. https://doi.org/10.1186/2041-1480-5-1.

Nicholls, J. G., & Paton, J. F. (2009). Brainstem: neural networks vital for life. Philosophical transactions of the Royal Society of London. *Series B, Biological sciences*, 364(1529), 2447–2451. https://doi.org/10.1098/rstb.2009.0064.

Niklison-Chirou, M., Agostini, M., Amelio, I., & Melino, G. (2020). Regulation of Adult Neurogenesis in Mammalian Brain. *International Journal of Molecular Sciences*, 21. https://doi.org/10.3390/ijms21144869.

Nishiyama, Y., & Katsura, K. (2015). *The Neuroendocrine System and Its Regulation*. 31-38. https://doi.org/10.1007/978-4-431-54490-6_3.

Nowrozi, M., & Goodarzi, N. (2022). Microanatomy of the cerebellum in Persian squirrel (Sciurus anomalus), a transmission electron microscopy, and immunohistochemical study. *Microscopy Research and Technique*, 85, 3850 - 3859. https://doi.org/10. 1002/jemt.24234.

O'Connor DC, Walker AE. Prologue. In: Walker Ae. *A History of Neurological Surgery*. New York. US: Hafner, 1967:1-22.

Obernier, K., & Álvarez-Buylla, A. (2019). Neural stem cells: origin, heterogeneity and regulation in the adult mammalian brain. *Development*, 146. https://doi.org/10.1242/ dev.156059.

Ortiz, C., Carlén, M., & Meletis, K. (2021). Spatial Transcriptomics: Molecular Maps of the Mammalian Brain. *Annual review of neuroscience*. https://doi.org/10.1146/ annurev-neuro-100520-082639.

Osanlouy, M., Bandrowski, A., De Bono, B., Brooks, D., Cassarà, A., Christie, R., Ebrahimi, N., Gillespie, T., Grethe, J., Guercio, L., Heal, M., Lin, M., Kuster, N., Martone, M., Neufeld, E., Nickerson, D., Soltani, E., Tappan, S., Wagenaar, J., Zhuang, K., & Hunter, P. (2021). The SPARC DRC: Building a Resource for the Autonomic Nervous System Community. *Frontiers in Physiology*, 12. https://doi.org/ 10.3389/fphys.2021.693735.

Paletzki, R., & Gerfen, C. (2019). Basic Neuroanatomical Methods. *Current Protocols in Neuroscience*, 90. https://doi.org/10.1002/cpns.84.

Paninski, L., & Cunningham, J. (2017). Neural data science: accelerating the experiment-analysis-theory cycle in large-scale neuroscience. *Current Opinion in Neurobiology*, 50, 232-241. https://doi.org/10.1101/196949.

Panova, E., & Lanska, D. (2020). Western European influence on the development of Russian neurology and psychiatry, part 1: Western European tours of early Russian neurologists and psychiatrists. *Journal of the History of the Neurosciences*, 30, 223 - 251. https://doi.org/10.1080/0964704X.2020.1840247.

Paridaen, J., & Huttner, W. (2014). Neurogenesis during development of the vertebrate central nervous system. *EMBO reports*, 15. https://doi.org/10.1002/embr.201438447.

Park, B. (2017). *Anatomy and Physiology of the Autonomic Nervous System*. 16, 101-107. https://doi.org/10.21790/RVS.2017.16.4.101.

Patton, P. (2015). Ludwig Edinger: The Vertebrate Series and Comparative Neuroanatomy. *Journal of the History of the Neurosciences*, 24, 26 - 57. https://doi.org/10.1080/0964704X.2014.917251.

Paulson, O., Schousboe, A., & Hultborn, H. (2023). The history of Danish neuroscience. *European Journal of Neuroscience*, 58. https://doi.org/10.1111/ejn.16062.

Pearson, K. (1993). Common principles of motor control in vertebrates and invertebrates. *Annual review of neuroscience*, 16, 265-97. https://doi.org/10.1146/ANNUREV.NE.16.030193.001405.

Peng, H., Xie, P., Liu, L., Kuang, X., Wang, Y., Qu, L., Gong, H., Jiang, S., Li, A., Ruan, Z., Ding, L., Yao, Z., Chen, C., Chen, M., Daigle, T., Dalley, R., Ding, Z., Duan, Y., Feiner, A., He, P., Hill, C., Hirokawa, K., Hong, G., Huang, L., Kebede, S., Kuo, H., Larsen, R., Lesnar, P., Li, L., Li, Q., Li, X., Li, Y., Li, Y., Liu, A., Lu, D., Mok, S., Ng, L., Nguyen, T., Ouyang, Q., Pan, J., Shen, E., Song, Y., Sunkin, S., Tasic, B., Veldman, M., Wakeman, W., Wan, W., Wang, P., Wang, Q., Wang, T., Wang, Y., Xiong, F., Xiong, W., Xu, W., Ye, M., Yin, L., Yu, Y., Yuan, J., Yuan, J., Yun, Z., Zeng, S., Zhang, S., Zhao, S., Zhao, Z., Zhou, Z., Huang, Z., Esposito, L., Hawrylycz, M., Sorensen, S., Yang, X., Zheng, Y., Gu, Z., Xie, W., Koch, C., Luo, Q., Harris, J., Wang, Y., & Zeng, H. (2021). Morphological diversity of single neurons in molecularly defined cell types. *Nature*, 598, 174 - 181. https://doi.org/10.1038/s41586-021-03941-1.

Perl, E. (2011). Pain and Nociception. *Comprehensive Physiology*, 915-975. https://doi.org/10.1002/CPHY.CP010320.

Pessoa, L., Medina, L., & Desfilis, E. (2021). Refocusing neuroscience: moving away from mental categories and towards complex behaviours. *Philosophical Transactions of the Royal Society B: Biological Sciences*, 377. https://doi.org/10.1098/rstb.2020.0534.

Pessoa, L., Medina, L., & Desfilis, E. (2021). Refocusing neuroscience: moving away from mental categories and towards complex behaviours. *Philosophical Transactions of the Royal Society B: Biological Sciences*, 377. https://doi.org/10.1098/rstb.2020.0534.

Piccolino, M. (1998). Animal electricity and the birth of electrophysiology: the legacy of Luigi Galvani. *Brain Research Bulletin*, 46, 381-407. https://doi.org/10.1016/S0361-9230(98)00026-4.

Pickersgill, J., Turco, C., Ramdeo, K., Rehsi, R., Foglia, S., & Nelson, A. (2022). The Combined Influences of Exercise, Diet and Sleep on Neuroplasticity. *Frontiers in Psychology*, 13. https://doi.org/10.3389/fpsyg.2022.831819.

Pirlot, P. (2015). Understanding Taxa by Comparing Brains. *Perspectives in Biology and Medicine*, 29, 499 - 509. https://doi.org/10.1353/PBM.1986.0019.

Polavaram, S., & Ascoli, G. (2017). An ontology-based search engine for digital reconstructions of neuronal morphology. *Brain Informatics*, 4, 123 - 134. https://doi.org/10.1007/s40708-017-0062-x.

Pretterklieber M. L. (2024). Nomina anatomica-unde venient et quo vaditis?. *Anatomical science international*, *99*(4), 333–347. https://doi.org/10.1007/s12565-024-00762-w

Price, R., & Duman, R. (2019). Neuroplasticity in cognitive and psychological mechanisms of depression: An integrative model. *Molecular psychiatry*, 25, 530 - 543. https://doi.org/10.1038/s41380-019-0615-x.

Puelles, Luis. (2012). *The Mouse Nervous System || Diencephalon*. 313–336. https://doi.org/10.1016/B978-0-12-369497-3.10009-3.

Raghanti, M., Munger, E., Wicinski, B., Butti, C., & Hof, P. (2017). *Comparative Structure of the Cerebral Cortex in Large Mammals*. 2, 267-289. https://doi.org/10.1016/B978-0-12-804042-3.00046-4.

Raj, B., Farrell, J., Liu, J., Kholtei, J., Carte, A., Acedo, J., Du, L., McKenna, A., Relić, Đ., Leslie, J., & Schier, A. (2020). Emergence of Neuronal Diversity during Vertebrate Brain Development. *Neuron*, 108, 1058-1074.e6. https://doi.org/10.1016/j.neuron.2020.09.023.

Reilly, C., & Fazekas, F. (2005). Communications of the European Neurological Society. *Journal of Neurology*, 252, 381-384. https://doi.org/10.1007/s00415-005-0843-7.

Rockland, K., & DeFelipe, J. (2016). Editorial: Neuroanatomy for the XXIst Century. Frontiers in Neuroanatomy, 10. https://doi.org/10.3389/fnana.2016.00070.

Roelfsema, P., & Holtmaat, A. (2018). Control of synaptic plasticity in deep cortical networks. *Nature Reviews Neuroscience*, 19, 166-180. https://doi.org/10.1038/nrn.2018.6.

Roeske, M., Konradi, C., Heckers, S., & Lewis, A. (2020). Hippocampal volume and hippocampal neuron density, number and size in schizophrenia: a systematic review and meta-analysis of postmortem studies. *Molecular psychiatry*, 26, 3524 - 3535. https://doi.org/10.1038/s41380-020-0853-y.

Rose, J., & Woodbury, C. (2008). *Animal Models of Nociception and Pain*. 333-339. https://doi.org/10.1007/978-1-59745-285-4_36.

Rushmore, R., Bouix, S., Kubicki, M., Rathi, Y., Yeterian, E., & Makris, N. (2020). How Human Is Human Connectional Neuroanatomy?. *Frontiers in Neuroanatomy*, 14. https://doi.org/10.3389/fnana.2020.00018.

Rushmore, R., Bouix, S., Kubicki, M., Rathi, Y., Yeterian, E., & Makris, N. (2022). HOA2.0-ComPaRe: A next generation Harvard-Oxford Atlas comparative parcellation reasoning method for human and macaque individual brain parcellation and atlases of the cerebral cortex. *Frontiers in Neuroanatomy*, 16. https://doi.org/10.3389/fnana.2022.1035420.

Sakai, T., Hata, J., Shintaku, Y., Ohta, H., Sogabe, K., Mori, S., Okano, H., Hamada, Y., Hirabayashi, T., Minamimoto, T., Sadato, N., Okano, H., & Oishi, K. (2020). The Japan Monkey Centre Primates Brain Imaging Repository of high-resolution postmortem magnetic resonance imaging: The second phase of the archive of digital records. *NeuroImage*, 273. https://doi.org/10.1016/j.neuroimage.2023.120096.

Sanchez, G., & Burridge, A. (2007). Decision making in head injury management in the Edwin Smith Papyrus. *Neurosurgical focus*, 23 1, E5. https://doi.org/10.3171/FOC.2007.23.1.5.

Sarter, M. (2004). Animal cognition: defining the issues. *Neuroscience & Biobehavioral Reviews*, 28, 645-650. https://doi.org/10.1016/j.neubiorev.2004.09.005.

Scarpazza, C., Scarpazza, C., Ha, M., Baecker, L., Garcia-Dias, R., Pinaya, W., Pinaya, W., Vieira, S., & Mechelli, A. (2020). Translating research findings into clinical practice: a systematic and critical review of neuroimaging-based clinical tools for brain disorders. *Translational Psychiatry*, 10. https://doi.org/10.1038/s41398-020-0798-6.

Schubiger, M., Fichtel, C., & Burkart, J. (2020). Validity of Cognitive Tests for Non-human Animals: Pitfalls and Prospects. *Frontiers in Psychology,* 11. https://doi.org/10.3389/fpsyg.2020.01835.

Schulte, B. (1986). *Neuroepidemiology as the Basis of Scientific Clinical Neurology*. 215-216. https://doi.org/10.1007/978-3-642-70007-1_27.

Serani-Merlo, A., Paz, R., & Castillo, A. (2005). The 'whole-animal approach' as a heuristic principle in neuroscience research. *Biological research*, 38 4, 359-64. https://doi.org/10.4067/S0716-97602005000400008.

Shepherd, G. (1972). The neuron doctrine: a revision of functional concepts. *The Yale Journal of Biology and Medicine*, 45, 584 - 599.

Shepherd, G., Marenco, L., Hines, M., Migliore, M., McDougal, R., Carnevale, N., Newton, A., Surles-Zeigler, M., & Ascoli, G. (2019). Neuron Names: A Gene- and Property-Based Name Format, With Special Reference to Cortical Neurons. *Frontiers in Neuroanatomy*, 13. https://doi.org/10.3389/fnana.2019.00025.

Shriver, A. J., & John, T. M. (2021). Neuroethics and Animals: Report and Recommendations From the University of Pennsylvania Animal Research Neuroethics Workshop. *ILAR journal*, 60(3), 424–433. https://doi.org/10.1093/ilar/ilab024.

Shterenshis, M., & Vaiman, M. (2007). European Influence on Russian Neurology in the Eighteenth and Nineteenth Centuries. *Journal of the History of the Neurosciences*, 16, 18 - 6. https://doi.org/10.1080/09647040500538125.

Siegel, R., & Watanabe, M. (2021). Introduction to the special focus on the development of the autonomic nervous system. *Birth Defects Research*, 113. https://doi.org/10.1002/bdr2.1902.

Simó, M., Navarro, X., Yuste, V., & Bruna, J. (2018). Autonomic nervous system and cancer. *Clinical Autonomic Research*, 28, 301-314. https://doi.org/10.1007/s10286-018-0523-1.

Sneddon, L. U., Elwood, R. W., Adamo, S. A., Leach, M. C. (2014). Defining and assessing animal pain. *Animal Behaviour*. 97:01-212. https://doi.org/10.1016/j.anbehav.2014.09.007.

Sneddon, L. (2015). Pain in aquatic animals. *The Journal of Experimental Biology*, 218, 967 - 976. https://doi.org/10.1242/jeb.088823.

Sneddon, L. (2018). Comparative Physiology of Nociception and Pain. *Physiology*, 33 1, 63-73. https://doi.org/10.1152/physiol.00022.2017.

Sneddon, L. (2019). Evolution of nociception and pain: evidence from fish models. *Philosophical Transactions of the Royal Society B*, 374. https://doi.org/10.1098/rstb.2019.0290.

Snyder, J. (2019). Recalibrating the Relevance of Adult Neurogenesis. *Trends in Neurosciences*, 42, 164-178. https://doi.org/10.1016/j.tins.2018.12.001.

Sotgiu, M., Mazzarello, V., Bandiera, P., Madeddu, R., Montella, A., & Moxham, B. (2019). Neuroanatomy, the Achille's Heel of Medical Students. A Systematic Analysis of Educational Strategies for the Teaching of Neuroanatomy. *Anatomical Sciences Education*, 13. https://doi.org/10.1002/ase.1866.

Spampinato, D., Casula, E., & Koch, G. (2023). The Cerebellum and the Motor Cortex: Multiple Networks Controlling Multiple Aspects of Behavior. *The Neuroscientist: a review journal bringing neurobiology, neurology and psychiatry,* 1073858 4231189435. https://doi.org/10.1177/10738584231189435.

Spencer, R., & Bland, S. (2019). *Hippocampus and Hippocampal Neurons. Stress: Physiology, Biochemistry, and Pathology.* https://doi.org/10.1016/B978-0-12-813146-6.00005-9.

Splavski, B., Rotim, K., Lakičević, G., Gienapp, A., Boop, F., & Arnautović, K. (2019). Andreas Vesalius, the predecessor of neurosurgery: How his progressive scientific achievements affected his professional life and destiny. *World neurosurgery*. https://doi.org/10.1016/j.wneu.2019.06.008.

Stahnisch, F. (2019). Catalyzing Neurophysiology: Jacques Loeb, the Stazione Zoologica di Napoli, and a Growing Network of Brain Scientists, 1900s–1930s. *Frontiers in Neuroanatomy*, 13. https://doi.org/10.3389/fnana.2019.00032.

Stephan, H., & Andy, O. (1969). Quantitative Comparative Neuroanatomy of Primates: An Attempt at a Phylogenetic Interpretation*. *Annals of the New York Academy of Sciences*, 167. https://doi.org/10.1111/j.1749-6632.1969.tb20457.x.

Stepien, B., & Wielockx, B. (2024). From Vessels to Neurons—The Role of Hypoxia Pathway Proteins in Embryonic Neurogenesis. *Cells*, 13. https://doi.org/10.3390/cells13070621.

Stiefel, M., Shaner, A., & Schaefer, S. (2006). The Edwin Smith Papyrus: The Birth of Analytical Thinking in Medicine and Otolaryngology. *The Laryngoscope*, 116. https://doi.org/10.1097/01.mlg.0000191461.08542.a3.

Stiles, J., & Jernigan, T. L. (2010). The Basics of Brain Development. *Neuropsychol Rev* 20, 327–348. https://doi.org/10.1007/s11065-010-9148-4.

Takemura, H., Pestilli, F., & Weiner, K. (2019). Comparative neuroanatomy: integrating classic and modern methods to understand association fibers connecting dorsal and ventral visual cortex. *Neuroscience research.* https://doi.org/10.1016/j.neures.2018.10.011.

Tobin, M., Musaraca, K., Disouky, A., Shetti, A., Bheri, A., Honer, W., Kim, N., Dawe, R., Bennett, D., Arfanakis, K., & Lazarov, O. (2019). Human Hippocampal Neurogenesis Persists in Aged Adults and Alzheimer's Disease Patients. *Cell stem cell*, 24 6, 974-982.e3. https://doi.org/10.1016/j.stem.2019.05.003.

Toda, T., Parylak, S., Linker, S., & Gage, F. (2018). The role of adult hippocampal neurogenesis in brain health and disease. *Molecular psychiatry*, 24, 67 - 87. https://doi.org/10.1038/s41380-018-0036-2.

Töppli, R.V. (1904). Anatomische Werke des Rhuphos und Galenos. *Erste Deutsche Übersetzung Anat H.* 25:342–472.

Tracey, D. (2017). Nociception. *Current Biology*, 27, R129-R133. https://doi.org/10.1016/j.cub.2017.01.037.

Vázquez-Guardado, A., Yang, Y., Bandodkar, A., & Rogers, J. (2020). Recent advances in neurotechnologies with broad potential for neuroscience research. *Nature Neuroscience*, 23, 1522 - 1536. https://doi.org/10.1038/s41593-020-00739-8.

Verma, K., & Kumar, S. (2022). How Do We Connect Brain Areas with Cognitive Functions? The Past, the Present and the Future. *NeuroSci*. https://doi.org/10.3390/neurosci3030037.

Walters, E. T., & Williams, A. C. C. (2019). Evolution of mechanisms and behaviour important for pain. Philosophical transactions of the Royal Society of London. *Series* B, *Biological sciences*, 374(1785), 20190275. https://doi.org/10.1098/rstb.2019.0275

Wang, L., & Orchard, J. (2019). Investigating the Evolution of a Neuroplasticity Network for Learning. *IEEE Transactions on Systems, Man, and Cybernetics: Systems*, 49, 2131-2143. https://doi.org/10.1109/TSMC.2017.2755066.

Wang, L., Liang, X., Wang, J., Zhang, Y., Fan, Z., Sun, T., Yu, X., Wu, D., & Wang, H. (2023). Cerebral dominance representation of directed connectivity within and between left-right hemispheres and frontal-posterior lobes in mild cognitive impairment. *Cerebral cortex*. https://doi.org/10.1093/cercor/bhad365.

Wang, Y., Guo, Y., Tang, C., Han, X., Xu, M., Sun, J., Zhao, Y., Zhang, Y., Wang, M., Cao, X., Zhu, X., & Guo, W. (2019). Developmental Cytoplasmic-to-Nuclear Translocation of RNA-Binding Protein HuR Is Required for Adult Neurogenesis. *Cell reports*, 29 10, 3101-3117.e7. https://doi.org/10.1016/j.celrep.2019.10.127.

Wallace, R., Olson, D., & Hooker, J. (2023). Neuroplasticity: The Continuum of Change. *ACS chemical neuroscience*. https://doi.org/10.1021/acschemneuro.3c00526.

Walsh, B., Attia, E., & Glasofer, D. (2020). Cognitive Neuroscience. *Eating Disorders*. https://doi.org/10.1093/wentk/9780190926595.003.0017.

Wehrwein, E., Orer, H., & Barman, S. (2016). Overview of the Anatomy, Physiology, and Pharmacology of the Autonomic Nervous System. *Comprehensive Physiology*, 6 3, 1239-78. https://doi.org/10.1002/cphy.c150037.

Wheeler, D., Banduri, S., Sankararaman, S., Vinay, S., & Ascoli, G. (2023). Unsupervised classification of brain-wide axons reveals neuronal projection blueprint. *Research Square*. https://doi.org/10.21203/rs.3.rs-3044664/v1.

White, C., Rees, C., Wheeler, D., Hamilton, D., & Ascoli, G. (2019). Molecular Expression Profiles of Morphologically Defined Hippocampal Neuron Types: Empirical Evidence and Relational Inferences. *Hippocampus*, 30, 472 - 487. https://doi.org/10.1002/hipo.23165.

Wick, Z., Tetzlaff, M., & Krook-Magnuson, E. (2019). Novel long-range inhibitory nNOS-expressing hippocampal cells. eLife, 8. https://doi.org/10.7554/eLife.46816.

Winnubst, J., Bas, E., Ferreira, T., Wu, Z., Economo, M., Edson, P., Arthur, B., Bruns, C., Rokicki, K., Schauder, D., Olbris, D., Murphy, S., Ackerman, D., Arshadi, C., Baldwin, P., Blake, R., Elsayed, A., Hasan, M., Ramirez, D., Santos, B., Weldon, M., Zafar, A., Dudman, J., Gerfen, C., Hantman, A., Korff, W., Sternson, S., Spruston, N., Svoboda, K., & Chandrashekar, J. (2019). Reconstruction of 1,000 Projection Neurons

Reveals New Cell Types and Organization of Long-Range Connectivity in the Mouse Brain. *Cell*, 179, 268-281.e13. https://doi.org/10.1101/537233.

Wm, C., Dh, H., & Er, K. (2000). The Emergence of Modern Neuroscience: Some Implications for Neurology and Psychiatry. *Annual Review of Neuroscience*, 23, 343-391. https://doi.org/10.1146/ANNUREV.NEURO.23.1.343.

Wolpe, P. (2002). The Neuroscience Revolution. (in Brief). *Hastings Center Report*, 32, 8.

Xi, W., Tian, M., & Zhang, H. (2011). Molecular imaging in neuroscience research with small-animal PET in rodents. *Neuroscience Research*, 70, 133-143. https://doi.org/10.1016/j.neures.2010.12.017.

Xu, N. (2020). Deciphering Pyramidal Neuron Diversity: Delineating Perceptual Functions of Projection-Defined Neuronal Types. *Neuron*, 105, 209-211. https://doi.org/10.1016/j.neuron.2019.12.018.

Xu, T., Yan, Y., Evins, A., Gong, Z., Jiang, L., Sun, H., Cai, L., Wang, H., Li, W., Lu, Y., Zhang, M., & Chen, J. (2020). Anterior Clinoidal Meningiomas: Meningeal Anatomical Considerations and Surgical Implications. *Frontiers in Oncology*, 10. https://doi.org/10.3389/fonc.2020.00634.

Yuste, R. (2015). From the neuron doctrine to neural networks. *Nature Reviews Neuroscience*, 16, 487-497. https://doi.org/10.1038/nrn3962.

Żakowski, W. (2020). Animal Use in Neurobiological Research. *Neuroscience*, 433, 1-10. https://doi.org/10.1016/j.neuroscience.2020.02.049.

Zargaran, A., Kordafshari, G., Hosseini, S., & Mehdizadeh, A. (2016). Akhawayni (?–983 AD): A Persian neuropsychiatrist in the early medieval era (9th–12th Century AD). *Journal of Medical Biography*, 24, 199 - 201. https://doi.org/10.1177/0967772014525105.

Zeng, H., & Sanes, J. (2017). Neuronal cell-type classification: challenges, opportunities and the path forward. *Nature Reviews Neuroscience*, 18, 530-546. https://doi.org/10.1038/nrn.2017.85.

Zentall, T. (2023). Comparative Cognition Research Demonstrates the Similarity between Humans and Other Animals. *Animals: an Open Access Journal from MDPI*, 13. https://doi.org/10.3390/ani13071165.

Zhang, X., Liu, L., Long, G., Jiang, J., & Liu, S. (2020). Episodic memory governs choices: An RNN-based reinforcement learning model for decision-making task. *Neural networks: the official journal of the International Neural Network Society*, 134, 1-10. https://doi.org/10.1016/j.neunet.2020.11.003.

Zheng, Y., Cong, X., Liu, H., Wang, Y., Storey, K., & Chen, M. (2022). Nervous System Development and Neuropeptides Characterization in Embryo and Larva: Insights from a Non-Chordate Deuterostome, the Sea Cucumber Apostichopus japonicus. *Biology*, 11. https://doi.org/10.3390/biology11101538.

Zhang, Z., Zhou, J., Tan, P., Pang, Y., Rivkin, A., Kirchgessner, M., Williams, E., Lee, C., Liu, H., Franklin, A., Miyazaki, P., Bartlett, A., Aldridge, A., Vu, M., Boggeman, L., Fitzpatrick, C., Nery, J., Castanon, R., Rashid, M., Jacobs, M., Ito-Cole, T., O'Connor, C., Pinto-Duartec, A., Dominguez, B., Smith, J., Niu, S., Lee, K., Jin, X., Mukamel, E., Behrens, M., Ecker, J., & Callaway, E. (2020). Epigenomic diversity of cortical projection neurons in the mouse brain. *Nature*, 598, 167 - 173. https://doi.org/10.1038/s41586-021-03223-w.

Zhou, M., Zhong, S., & Verkhratsky, A. (2023). Astrocyte syncytium: from neonatal genesis to aging degeneration. *Neural Regeneration Research*, 19, 395 - 396. https://doi.org/10.4103/1673-5374.379047.

Zoccoli, G., & Amici, R. (2020). Sleep and autonomic nervous system. *Current Opinion in Physiology*, 15, 128-133. https://doi.org/10.1016/j.cophys.2020.01.002.

About the Authors

Tantri Dyah Whidi Palupi
Lecturer of the Veterinary Anatomy Division
Faculty of Veterinary Medicine, Universitas Airlangga (UNAIR), Indonesia

Tantri Dyah Whidi Palupi, drh., MSi, was born in Madiun, Indonesia, on August 2, 1997. She completed her Doctor of Veterinary Medicine (Drh.) and Master of Science (MSi) in Reproductive Biology at the Faculty of Veterinary Medicine, Universitas Airlangga, where she also earned her bachelor's degree. Dr. Tantri is deeply passionate about animal macro-anatomy and micro-anatomy. Currently, she focuses on her expertise on animal anatomy, orthopedics in animals, and reproductive neurohormones. Her research interest includes the morphology and ultrastructure of animal reproductive organs, reproductive disorders in animals, as well as neuroanatomy and neurohormonal studies.

Soeharsono
Lecturer of the Veterinary Anatomy Division
Faculty of Veterinary Medicine, Universitas Airlangga (UNAIR), Indonesia

Dr. Soeharsono, drh., MSi, was born in Tulungagung on April 2, 1961. He completed his bachelor's degree at the Faculty of Veterinary Medicine, Universitas Airlangga, followed by a master's degree from IPB University and a doctoral degree from Universitas Brawijaya. Dr. Soeharsono has a profound interest in veterinary anatomy. He currently resides in Driyorejo, Gresik, East Java, where he continues to contribute to the field of veterinary sciences.

Yeni Dhamayanti
Lecturer of the Veterinary Anatomy Division
Faculty of Veterinary Medicine, Universitas Airlangga (UNAIR), Indonesia

Dr. Yeni Dhamayanti, drh., MKes, was born in Bandung on March 3, 1967. She completed her bachelor's degree at the Faculty of Veterinary Medicine, Universitas Airlangga, where she later obtained her Doctor of Veterinary Medicine. Dr. Yeni also holds a Master of Health in Basic Medicine Science from the Faculty of Medicine, Universitas Airlangga, and a Doctoral degree in Science from the Graduate Program at Universitas Airlangga. Dr. Yeni has a strong interest in Photogrammetry and has developed expertise in Animal Anatomy. Her research interests include Photogrammetry, Sex Bird Dimorphism, and Laserpuncture.

Gracia Angelina Hendarti
Lecturer of the Veterinary Anatomy Division
Faculty of Veterinary Medicine, Universitas Airlangga (UNAIR), Indonesia

Gracia Angelina Hendarti, drh., MSi was born in Pangkal Pinang City, province of Bangka Belitung, on September 13th, 1970. She studied at the Faculty of Veterinary Madicine, Airlangga University, Surabaya, and graduated in 1996. Then, she continued her Master's degree in Reproductive Biology at the Universitas Airlangga Postgraduate Program. Currently, she is taking a Doctoral degree and working as a lecturer in Veterinary Anatomy Division at the Faculty of Veterinary Medicine, Airlangga University Surabaya since 1999. Gracia has an interest in Functional Anatomy and Anatomy Physiology.

Index

F

G

H

I

L

M

N

P

R

S

T